SYSTEMS BIOLOGY – THEORY, TECHNIQUES AND APPLICATIONS

THE MECHANICS OF LIFE

A CLOSER LOOK AT THE INNER WORKINGS OF NATURE

Systems Biology – Theory, Techniques and Applications

Additional books in this series can be found on Nova's website under the Series tab.

Additional e-books in this series can be found on Nova's website under the eBooks tab.

SYSTEMS BIOLOGY – THEORY, TECHNIQUES AND APPLICATIONS

THE MECHANICS OF LIFE

A CLOSER LOOK AT THE INNER WORKINGS OF NATURE

TIMOTHY GANESAN

nova science publishers
New York

Library of Congress Cataloging-in-Publication Data

ISBN: 978-1-53612-937-3

Published by Nova Science Publishers, Inc. † New York

CONTENTS

PREFACE

"What gets us into trouble is not what we don't know, it's what we know for sure that just ain't so."

– Mark Twain

In the past couple of years, biology and medicine have significantly evolved. Technological developments in robotics, computing and quantum physics have spilled over into many disciplines including the life sciences. The idea of the solitary biologist taking samples from the field and working in a lab with test tubes and chemical apparatus is a thing of the past. Today, bio labs are filled with scientists across various disciplines like physics, mathematics and engineering. Most equipment are completely automated and robotic. They are often controlled by powerful computers acquiring vast amounts of data for rigorous analysis. These advances have sped up the rate of research done in biology and medicine. Although the tech revolution in biology has been going on for some time now, things really started taking a turn in 2010 when the geneticist, Craig Venter created the world's first synthetic life form. His team consisted of 20 scientists working close to over a decade. Their target was to completely engineer an organism from scratch. They believed that if they could do this, then the sky was the limit. They could then design organisms that churn out biofuels or manufacture vaccines. Basically they can make living things do whatever they wanted. But to achieve this feat, Venter's team had to fuse various scientific disciplines – from the technology they used to the powerful computers needed for simulation. They did just this. Successfully engineered, they christened their artificial bacteria, *Mycoplasma Laboratorium*.

The availability of powerful computers have enabled scientists to significantly cut down experiments, expediting research. Venter's research would have taken several decades if not for cutting-edge computing and robotics available to them. Speaking of robots, they too have found their way into the life sciences. In medicine, robots are already being used for certain types of surgery and therapies. In 2014, Australia's Science Agency collaborated with India to explore the depths of the Indian Ocean (Davidson, 2014). This research focused on what goes on in the depths of the Indian Ocean. To study this, the teams released underwater robots that go to different depths to take measurements. The data was then sent back data to the surface. The robots gathered critical data on plankton populations. Being food for most marine creatures, plankton are seated at the base of the food chain. Such data gives us a good idea of the condition of the ecosystem. And a good understanding of the ecology is key to conservation. Towards such efforts, ecologists are heavily relying on computer programs to simulate whole ecosystems.

Besides robotics and computers, the recent years also saw significant strides in genetic engineering. The most effective technique currently used for gene editing is CRISPR. The method allows for molecular-level accuracy for manipulating genes. Using CRISPR, scientists are even thinking about stopping entire epidemics. This is biology today; a mixture of techniques from across various fields. Along with these advances, this book also touches on issues related to disease spread, medical efforts and scientific practices. When we talk about advances in biology, it's only natural that we stumble upon evolution; the fundamental theory of life. We see the mechanics of how bacteria evolve antibiotic resistance and why parasites are so complicated. Reviewing the evolution of life on our planet, ve see our own evolution – the evolution of Homo sapiens. In essence, we ake a closer look at the Mechanics of Life.

References

Davidson, H., (2014), Scientists to launch bio robots in Indian Ocean to study its 'interior biology, *The Guardian*: https://www.theguardian.com/environment/2014/jan/30/scientists-launch-bio-robots-indian-ocean-study-internal-biology

ACKNOWLEDGMENTS

I am most thankful to my grandmother and aunts; they seem to never run out of support and wisdom. I am extremely grateful to constantly have the following people in my life: Sima Alian, Suren Lim Sinnadurai, Manoj Joseph Matthew, Noraz Khairi, Mohammad Nazmi Zaidi Moni, Deep Kumar, Irraivan Elamvazuthi, Pandian Vasant, Ajitpal Singh Sekhon, Haravinthan Arumugam, Ahmad Zulazlan Shah Zulkifli and Mohammad Shiraz Aris. Many thanks to my editors at Nova Science Publishers especially Nadya S. Gotsiridze-Columbus; who took up this project.

Chapter 1

PARASITES: COMPLEXITY BEYOND EVOLUTION

"Nothing in life is to be feared, it is only to be understood. Now is the time to understand more, so that we may fear less."

— Marie Curie

IN SEARCH OF MICROBES

As opposed to symbiotic life forms, parasites are organisms that have non-mutual benefit relationships with their hosts. For instance: the group of plants known as Mistletoe absorbs food and water from various types of trees (e.g., apple tree) without giving any benefit to the host plant. The human body itself contains approximately 300 species of parasitic worms and 70 species of protozoa (which are the same of type of microbes that cause malaria) (Ashford and Crewe, 1998). Interestingly only 10 percent of our cells in the human body are truly ours, the remaining 90 percent belong to bacteria and parasites (Buckman, 2002). Among these parasites, a small group of them are known to cause some of the most widely spread diseases in the world such as malaria and tape worm infection. Since parasites are often found in the tropics, the science of parasites has been in the domain of tropical medicine. Today it has become known that a large portion of organisms on our planet are parasites. These parasites are in constant interaction with their host(s) and other parasites. Therefore, the study of parasites has spilled over from tropical medicine into the vast fields of microbiology, ecology and even behavioral studies (e.g., study of human interaction and disease spread). Even new fields of research such as paleoparasitology have emerged – where scientists study parasitic infections in ancient societies from archeological artifacts.

Among the earliest records of parasitic infections were found from Egyptian manuscripts (3000 to 400 BC) as well as in the works of the great Greek physician Hippocrates (800 to 300 BC) (Bryan, 1930). Towards the turn of the first millennium (about 900 AD) Persian physicians, Rhazes and Avicenna made significant strides in medicine providing clear descriptions of many parasite-causing diseases. These physicians recognized and characterized various types of parasitic worms (or helminths) especially in the regions surrounding the Red Sea. Such helminths usually colonize the intestinal tracts of humans e.g., pin worms, tape worms and guinea worms.

Modern day scientific knowledge on parasites owes much to the development of germ theory and bacteriology. Among the luminaries known for the development of these ideas were Louise Pasteur and Robert Koch. Pasteur was a French chemist who studied in the Ecole Normale in Paris and later on did research at University of Lille (Louis Pasteur, 2016). While at the university, Pasteur was approached by an employee from the brewery that made beer from sugar beets. The employee explained to Pasteur the problem faced in the brewery where many barrels of beer were turning bad and thus had to be thrown away. This issue was causing the brewery to lose a lot of money. The employee asked Pasteur to take a look at the problem and see if he could help in any way. Pasteur proceeded to take samples from the beer barrels and studied them under a microscope. He found that the samples contained thousands of microbes. Being a proponent of germ theory, Pasteur became convinced that the microbes were causing the beer to go bad. Germ theory states that diseases are caused by microbes, as opposed to the theory of spontaneous generation. The theory of spontaneous generation claims that organisms (or diseaseausing microbes) could spontaneously materialize from dust or rotting lesh. Although you may think spontaneous generation to be ridiculous at irst glance, this theory held sway for two thousand years unchallenged. Jpon finding these microbes, Pasteur tenaciously continued his research nto various liquids (such as milk, wine and vinegar) albeit being ridiculed by his colleagues in medical institutions. Pasteur then invented a witty way not only to kill the microbes in the liquid sample but also to prolong the time it takes for the liquid to go bad. This process is known today as Pasteurization and has been successfully employed in various milk, wine and beer production industries. This process can be carried out simply by heating the liquid to kill most of the microbes and then avoiding air (which is the source of the microbes) from getting into it and thus prolonging its shelf-life. In these experiments Pasteur explains that the bacteria from air cause the liquid to go bad and not the air itself. The curved neck flask (known today as the 'Pasteur flask') showed in Figure 1.1 was employed by Pasteur to avoid air from entering the flask and reaching the sample.

Around 1865, Pasteur was approached by the silk industry to help find the cause for a disease called Pebrine – which infects silk worms. In his investigations, Pasteur found that the disease was caused by another organism. Today it is well known that Pebrine is caused by some fungal parasites. Pebrine causes brown dots to form on the infected silkworm larvae. The infected larva then becomes incapable of spinning silk thread. After this business with the silk industry, Pasteur becomes convinced that microbes could also spread disease among humans. While Pasteur was developing methods in the spirit of germ theory in France, an English physician Edward Jenner was pondering about the mechanisms at which smallpox propagates. Edward Jenner was a country doctor who was very much fascinated with nature. He was particularly interested in the lore that milkmaids are immune to smallpox. Jenner believed that instead of smallpox, milkmaids caught a weaker derivative called cowpox. He conjectured that the pus from the blisters of the cowpox shields the milkmaids from contracting smallpox. The lack on regulations at that time gave Jenner the questionable freedom to do some human testing to assess his hypothesis. Hence. Jenner found himself a test subject – a boy called

James Phipps. He then extracted pus from the blisters of a cowpox-infected milkmaid.

The cowpox pus was then injected into the boy repeatedly for a couple of days – gradually increasing the amount of the injection. Then Jenner deliberately injected the boy with smallpox. It was observed that the boy became slightly sick but quickly had a complete recovery in a few days with no adverse effects. Similar to Pasteur, Jenner faced serious criticism the medical community and other scientists (and not because of his scientific ethics). In London, Jenner was publicly humiliated since they could not come to terms with the fact that a country doctor such as Jenner could have made such a monumental discovery. Gradually the effectiveness of Jenner's method was undeniable and it became the standard treatment for smallpox. Jenner considered his treatment method as a gift to the world. Thus he did not patent it since he reasoned that it would only make the treatment more expensive and inaccessible to those who cannot afford it. Jenner's method is today known as vaccination. In 1980, the WHO declared that global extinction of smallpox.

Pasteur came to know about the Jenner's vaccination methods and thought that such a vaccine could be developed for all the diseases in the world. In that time Pasteur formed a research group to study chicken cholera. Chicken cholera is a disease that is caused by bacterial infection. Although it mainly affects chickens, this type of cholera could also be transmitted to humans. In 1880, using the vaccination method of Jenner, Pasteur injects some chickens with chicken cholera germs from an old culture. Due to the weakness of the germs, the chickens did not die. As a control experiment, he injected another group of chickens with germs from a new culture which resulted in the death of the chickens. Then he observed that the chickens injected with the old germ culture were immune to chicken cholera. Thus Pasteur and his research team developed a vaccine for chicken cholera.

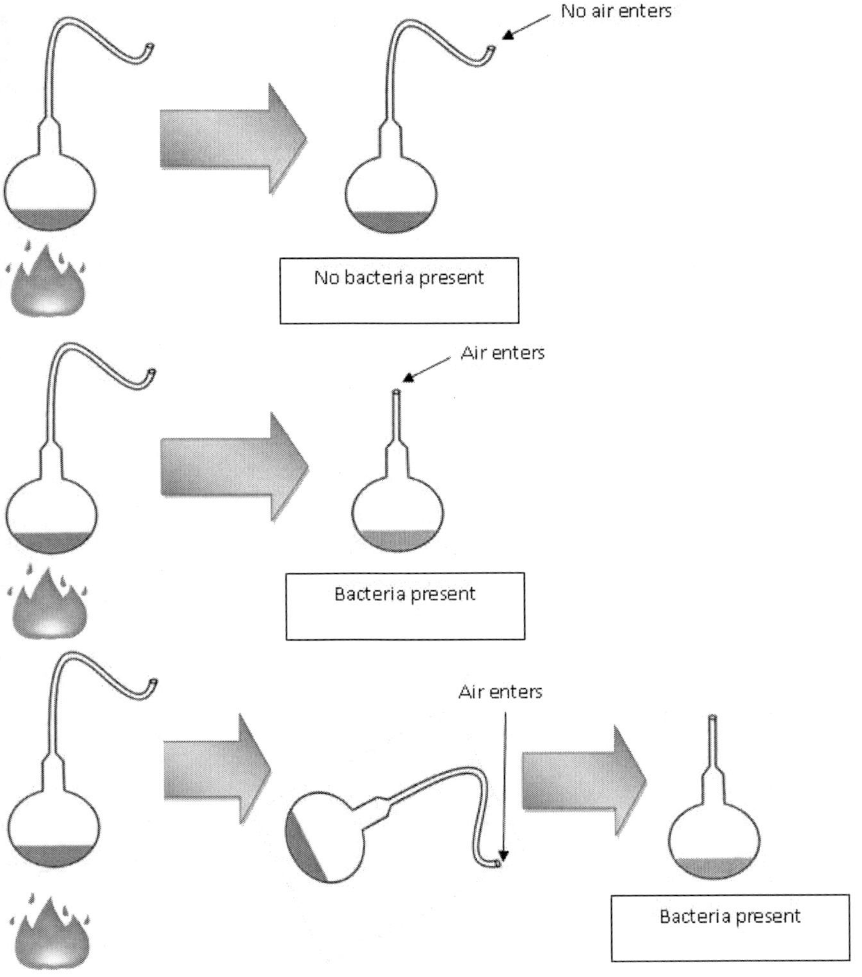

Figure 1.1. Pasteur's Experiment – Evidence for Germ Theory.

Despite various rejections and mockery by the scientific community, Pasteur developed a vaccine for anthrax using the same method the following year. Similarly, anthrax is caused by bacteria which are transmitted via contaminated meat affecting most animals including humans. The efficacy of Pasteur's anthrax vaccine was even challenged by the editor of the "The Veterinary Press". Pasteur rose up to the challenge and performed the vaccination experiment on a group of sheep. The experiment was a success proving that the inoculated sheep was immune to anthrax and did not suffer any of its symptoms. Armed with the success of chicken cholera and anthrax, Pasteur's research team now turned their attention to rabies – which was becoming more common in France. Rabies is a viral disease transferrable to humans or animals through the bite of a rabid animal. Rabies usually occurs in animals such as raccoons, dogs, skunks, bats and foxes affecting the central nervous system resulting in brain disease and death (Rabies, 2016). With a slight modification to the vaccination strategy, Pasteur managed to successfully vaccinate animals and expand his vaccine for implementation on humans. Armed with Jenner's methodology of vaccination, Pasteur and his team developed vaccines for chicken cholera, anthrax and rabies.

While Jenner and Pasteur developed and enhanced the vaccination method, Robert Koch a German physician was trying to provide concrete evidence to germ theory by finding the microbes responsible for specific diseases (Koch, 2016). Pasteur's work on cholera was not successful where he was not able to identify the associated microbe. Unlike Pasteur, Koch had a strong foundation of the human anatomy and his scientific rigor was commendable – he could patiently work tirelessly for long hours. Koch then started to study microbes in his surgical laboratory. During this time Koch was a district medical officer near Berlin, Germany. Using his laboratory, Koch researched Anthrax which was at the time plaguing farms and taking the lives of many herd animals. For three years Koch studied about the characteristics of anthrax from its symptoms in animals to its life cycle. Koch found out that the anthrax microbe (since they were aware of the existence of bacteria at the time) produced spores that stayed on the farm animal long after its death. This spore then has the capability to transmit the disease to other animals. Koch continued to study the diseases that specifically infected humans such as septicaemia or blood poisoning. Septicaemia is caused by a localized infection which spreads into the blood and causes other systems in the body to become infected. Koch conducted research to identify the microbe that was causing septicaemia. Analyzing the blood sample under the microscope, Koch could not see its presence.

In the 1870s other German researchers such as Karl Weigert and Paul Ehrlich were busy experimenting with dyes to stain specific bacteria.

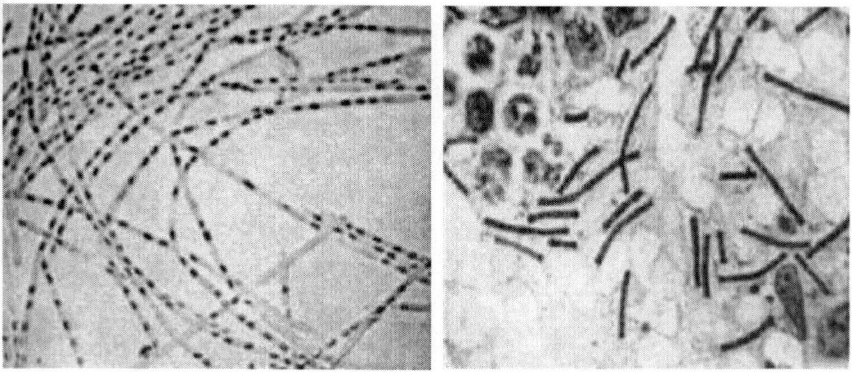

Figure 1.2. The first photographs of the bacteria, anthrax bacillus (Source: Koch, 1877).

Influenced by these scientists Koch then tried dying septicaemia-infected blood samples. He discovered that methyl violet dye showed the septicaemia germ under the microscope. Photographing the dyed germ, Koch enabled people from outside his laboratory to view the germ (Blevins and Bronze, 2010). Similarly using dyes, Koch managed to identify and photograph the anthrax microbe (see Figure 1.2).

Koch even improved the experimental technique for growing cultures of germs in labs using a mixture of potatoes and gelatine. By the use of this approach germs could be studied better. Similar to Pasteur, Koch then assembled a team of scientists and began to study one of the most dangerous diseases of that time, tuberculosis (TB). Using his dye-staining method, Koch and his team analyzed the TB infected samples. They observed the presence of rod-shaped bacteria (tubercle bacteria) in the infected lesions (Cambau and Drancourt, 2014). These tubercle bacteria looked very similar to *Mycobacterium leprae* – the bacteria that causes leprosy. Noticing this, they isolated the tubercle bacteria using a novel solid culture media approach. Koch was then awarded the Nobel Prize of Medicine in 1905 for his work on tuberculosis.

Besides constructing conclusive proof of germ theory, Koch also showed that specific microbes could be causally linked to the onset of certain diseases. Along the way Koch also invented critical methodology for observing and isolating microbes opening up a whole new field of biological and medical research known today as bacteriology and microbiology. Most infections and diseases studied by them were caused by microbes (bacteria and viruses) which were parasites. The genius of Jenner's vaccination approach, Pasteur's inventions of various vaccines along with Koch's pioneering methods of microbe identification revolutionized our thinking of modern medicine and biology.

A Parasitic Beginning

Due to the large size of parasitic worms (helminths), humans were aware of their existence since the earliest times. The primitive tribes of North Borneo have been known to be aware of their intestinal roundworms and tapeworms (Hoeppli, 1959). In Greece, Hippocrates discovered helminths in humans and other organisms (e.g., fishes and domesticated animals). Roundworms are also known as nematodes are a group of worms inhabiting a vast range of environments. More than half of the nematodes known are parasites which live off plants, fungi, bacteria, large animals and unfortunately humans. Nematodes are highly resilient parasites and have developed various adaptations to adverse environmental conditions. When nutrients (or water) become scarce or temperatures become extreme, nematodes could initiate what is called 'cryptobiosis'. During cryptobiosis, the nematode stops all metabolic or energy consuming activities (e.g., reproduction, development or self-repair) and goes into a death-like state. Nematodes could maintain this state indefinitely until the environment becomes habitable again. They are pretty tough worms.

One common parasitic nematode is *Ascaris lumbricoides* tapeworm which causes a disease called Ascariasis. The symptoms of Ascariasis are; fever followed by abdominal swellings and diarrhea. The Ascaris worm has been found in fossilized human feces in Peru and Brazil. And records show that there were traces of the worm in Egyptian mummies and Ming Dynasty remains in China (Horne, 1985; Ferreira et al., 1983; Dexiang et al., 1981). Globally, about one billion people are infected by the ascaris worm. This worm lives mostly in the intestines where it matures and lays ggs. As disgusting as this may sound – the eggs are passed out through ne feces and transmitted when food gets contaminated with the eggs.

One of the discoverers of the worm's transmission method was the talian physician Giovanni Battista Grassi (Cox, 2002; Boese, 2011). Grassi found *ascaris* worms (along with eggs) in the large intestines of a body during an autopsy procedure. This was 1878, about the same time germ theory was being solidified by Koch and Pasteur. To prove that the ascaris worm infected human hosts via the ingestion of the worm's eggs, Grassi carried out a very disgusting but critical experiment – he ate the worm's eggs extracted from the corpse's intestines (sorry I just had to rush to the bottom line). However, before Grassi consumed the eggs, he had to ensure that he was not already infected. Not only was Grassi's experiment repulsive, it was also very lengthy – where for a whole year Grassi examined his feces daily with a microscope to see if there were any traces the *ascaris* tapeworm or its eggs. Once he was sure he was not infected, he extracted and consumed the eggs as mentioned above. Grassi's experiment showed positive results a month later when he identified tapeworm eggs in his stool. He treated himself with herbal tapeworm medicine to remove the parasites from his intestines. Thanks to physicians like Grassi, we now have a better understanding of the life cycle and transmission modes of helminths such as the ascaris tapeworm.

Following closely the path laid by Grassi, Shimesu Koino a Japanese pediatrician attempted to study the migration path of the *ascaris* tapeworm in the human body during infection. Therefore, similar to Grassi, he ate about 2000 tapeworm eggs, while his brother consumed 500 of the mentioned eggs. These eggs were extracted from pigs. Consequently, Koino developed pulmonary syndrome (or pneumonia-like breathing condition). He then identified traces of tapeworm larvae in his coughed up mucus ascertaining the migration route of the tapeworm. Thus Koino's experiment established two critical facts: the ascaris tapeworm infection causes pneumonia like syndrome at the initial stages (four days to two weeks) of the disease. Second, the tapeworm larva migrates from the lungs to the stomach (Koino, 1922). Along with these discoveries, Koino and his brother's health recovered and there were no lasting effects.

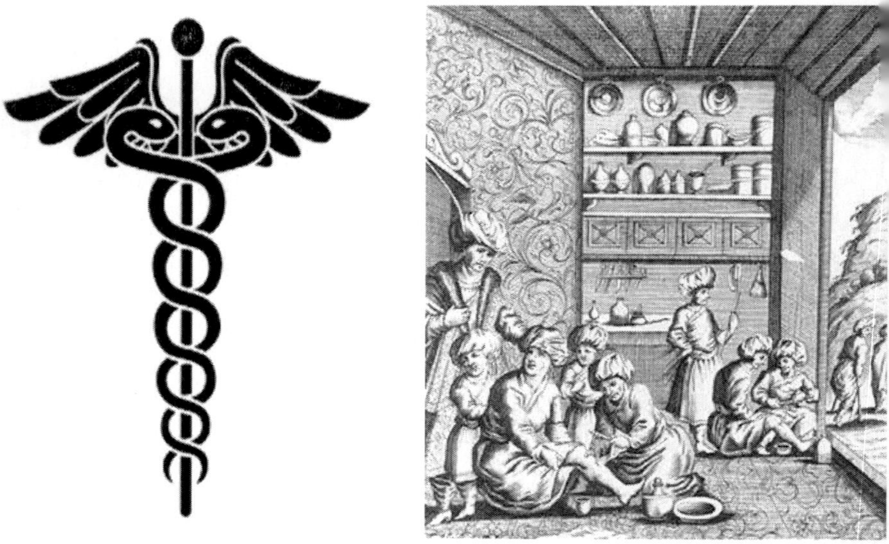

Figure 1.3. International Symbol of Medicine (Left) and Depiction of Persian physicians winding the guinea worm on a stick to remove it from the patient's leg (Right).

Another type of helminth known to infect humans is the hookworm. Hookworm infection causes anemia and a greenish yellow paleness on the patient. The Trichinella spiralis worm is one other parasite that causes *trichinellosis* disease. Most of its infections in humans are caused by the consumption of infected pork. The trichinella worm infects various mammals such as swine, horses and other wild animals. It is highly possible that the Islamic and Judaic practice of avoiding the consumption of pork may have been due to the infections of these worms – since it is highly prevalent in swine (Cox, 2002). Trichinellosis disease could be avoided by consuming well-cooked pork – since extensive heating kills the Trichinella spiralis worm. Guinea worms also known as *Dracunculus Medinensis* (causing Guinea worm disease) is a parasite first mentioned in documents written in 1500 BC. Traces of these worms have also been found in ancient Egyptian mummies – sometimes called the Pharaoh's Worm.

Dracunculus Medinensis has been mentioned in the Bible as 'fiery serpents' that struck the Israelites in the region surrounding the Red Sea during the Exodus from Egypt. Beside Greek and Roman texts dating 1000 and 1100 AD, other references to Guinea worms include a 700 BC Assyrian text owned by King Ashurbanipal (History of Discovery, 2006). In that same Assyrian text it was stated that the disease was probably brought to Egypt by Mesopotamian prisoners. The Persian physicians, Rhazes and Avinenna significantly contributed to our understanding of the Guinea worm disease which they termed 'Medina vein' and the Guinea worm 'The Little Dragon of Medina'. These physicians were treating the infections by winding the worms on sticks to remove them from the patient's body (right – Figure 1.3). It has been speculated that the symbol of medicine (left – Figure 1.3) symbolically represents this ancient medical practice of worm winding. Unfortunately there is still no drug which can prevent or cure Guinea worm disease. The up-side is Guinea worm disease could be easily prevented simply by taking containment measures, ensuring safe water supply (e.g., using filters), disinfecting the water source/supply and frequent monitoring of infection level. For most helminths infections (which are soil-transmitted), the WHO recommends periodic medical treatment (also known as deworming) for those who live in high-risk areas. Besides it also recommends health and hygiene education in addition to access to adequate sanitation in impoverished settings – to prevent parasite infestation and soil-contamination with infected worm eggs. This is because the Ascaris worm and hook worm lay eggs that become present in human feces. The worms then spread to the soil and propagate – especially in areas where sanitation is poor.

The invention and usage of the microscope by Anton Von Leeuwenhuk stimulated research in microorganisms – any organisms which are microscopic in size. These research efforts have shed some light to our understanding of various parasitic microbes. Microbes, parasitic or not live in every segment of our biosphere – which includes hot springs, soil, high up in the atmosphere and deep down under the ocean. Microbes are highly resilient to exterior conditions and can thus thrive in very extreme environments. In 2013, an international team of scientists discovered that there were many active microbes deep in the Mariana Trench. The microbes were discovered at about 11, 000 meters below sea level which is one of the most inaccessible areas on earth (Glud et al., 2013). Being the deepest places on the planet, this extreme site harbors microbes at pressures up to 1100 times more than at sea level. Deep sea trenches such as this provide a suitable deposit for organic matter (e.g., fish carcasses, algae, etc.). Feeding on organic matter, the microbes create an ecosystem with other microbes as well as small organisms which are similarly evolved to live in extreme environmental pressures. The team of scientists made the discovery by deploying a lander into the trench. Remotely controlled from the ship, the lander obtained samples of unique data.

Microbes could be grouped based on their characteristics e.g., bacteria, protozoa, viruses, fungi and archaea. Our environment is teeming with microbes – parasitic microbes are merely a fraction of them and the ones that may cause harm to humans are a fraction of that fraction. A group of microbiologists estimate that if a single bacterium is represented by a penny, then the total amount of stacked pennies would reach a distance of a trillion light years (a trillion refers to one with twelve zeroes behind it). Hence the distance achieved by the stacked pennies would take a trillion years to cross if one were to be travelling at the speed of light – and that is a rough measure of how much of bacteria exists on earth. It was also found that most bacteria live in soil and oceans – including their subsurfaces. Like most life forms, bacteria contain carbon which is employed for various biological processes. Carbon takes up one half of a bacterium's dry weight. The group of microbiologists found that the total bacterial carbon found in soil subsurface on earth is equivalent to the approximate weight of the United Kingdom (Whitman et al., 1998).

The human intestine is a habitat for nine species of amoeba – a protozoa-type microbe. Similar to helminths and bacteria, certain types of protozoa are parasites and may cause disease like malaria, sleeping sickness and dysentery. Of these nine amoeba species in the human gut, only one of them is pathogenic and can cause harm – *E. histolytica*. Similar to helminths, amoeba live and breed in the intestines and transmitted out in the feces – the consumption of food or water contaminated with these amoebas enables it to infect new hosts. The most eerie parasitic amoeba is the *amoeba naegleria*. This species of amoeba lives in warm fresh water such as lakes and rivers – it enters humans via the nasal passage and fatally infects and destroys brain tissue eventually causing death.

HOST CONTROL

Since the very beginnings of civilization, humans have been aware of the existence of parasites – especially the ones that infect us. Their early exploration of parasites covered: disease descriptions, parasites identification, parasite-disease causal connections and parasite life cycles. Today scientists have discovered that the life cycles of some parasites are far more complex than our initial expectations. As newer discoveries are made, our initial ideas on their role in the ecosystem have come to close scrutiny. One interesting phenomena in parasitic life cycles is host manipulation – where the parasite manipulates the host to achieve its own goal(s). Such parasites have been observed to manipulate their hosts behavior and/or characteristics (known as phenotypes) (Poulin, 1998). For instance; there is a species of hairworm (*Spinochordodes tellinii*) known to infect grasshoppers and crickets. The adult hairworm lives and breeds in water. The crickets and grasshoppers that drink this water ingest the hairworm larvae which then develop in the host's body. Once completely developed, the hairworm releases a chemical that takes over the central nervous system of the host. This chemical alters the behavior of the insect such that it finds for a body of water and jumps into it so that the parasite hairworm is released into the water for it to continue living and breeding repeating the cycle. The grasshopper or cricket drowns and dies during the process of releasing the hairworm. The hairworm thus makes the host suicidal for its own benefit.

Another peculiar parasite is the mermethid worm. One of the mermethid worm's hosts is the mayfly. The adult mermethid lives freely in freshwater systems unlike their young which live in other species of insects like the mayfly. Young or immature mayflies have gills and are completely aquatic – known as nymphs. The mayfly nymphs stay in that state for about a year before growing to an adult. The mayflies remain in adult form for a maximum of two days at during which they mate with other adults. After mating the females leave their eggs in the water while the males die. The mayfly nymphs eat the eggs of the mermethid worm while in the water. The nymphs then carry the mermethid worm parasite throughout their lives. In a female mayfly, the mermethid worm grows in her body until she goes to deposit her eggs. Once she deposits her eggs, the mermethid worm breaks free from her abdomen and returns into the water. However, when a mermethid worm infects a male mayfly returning back into the water may be a bit of problem – since the males don't lay eggs. The worm thus interferes with the hormonal system of the male mayfly – completely reversing its sexuality making it female. One way or another, the mermethid worm has its way. Consequently the feminized mayfly mates with other males before proceeding to the water – where the mermethid worm frees itself (Vance, 1996). This parasite not only feminizes the physiology of the male mayfly but also alters its behavior such that it mates with other males.

Similarly there is a species of wasp (*Reclinervellus nielseni*) which parasites on the spider to aid in its care for its offspring. The uninfected host spider (*Cyclosa argenteoalba*) is capable of weaving two types of web structures; regular and resting web. The regular web structure is weaved by the spider to trap insects and other organisms for food. The resting web structure is employed by the spider when it molts. Spiders like other invertebrates have a skeletal structure which is outside of its body – an exoskeleton. So at some time during its development, spiders outgrow their exoskeleton and generate a new one. During this time it would have to leave the previous exoskeleton, not unlike a snake shedding its old skin – this transition is called molting. The resting web structure is weaved by the spider for molting. The female parasitic wasp initially infects the host spider by attacking it and injecting its eggs in the spider's abdomen. Once hatched the larvae survives in the spider by feeding on spider blood also known as hemolymph. During the growth of the larvae in the spider, it begins to release chemicals mimicking the spider's molting hormones. These hormones cause the spider to begin building its molting web as though it is preparing to molt. Once the structure is complete, the larvae leave the spider – killing it in the process. The larvae then spin the spider's molting web forming it into a cocoon where it lives while it further develops into an adult wasp. The adult wasp then mates and the process repeats (Ross, 2015; Takasuka et al., 2015).

For the purpose of manipulating their hosts the hairworm hijacks the central nervous system of the grasshopper (or cricket) while the parasitic wasp and the mermethid worm interferes with the hosts hormones.

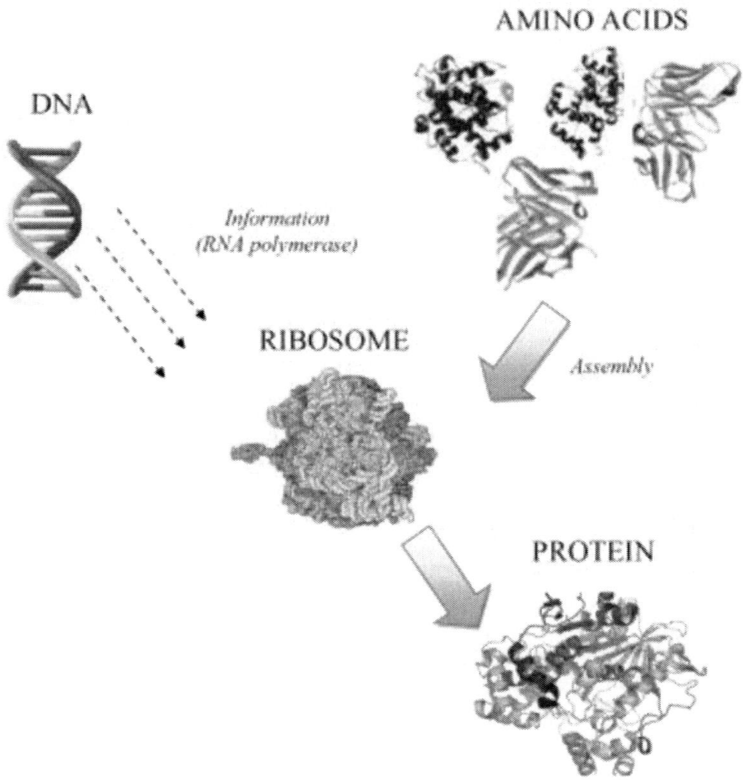

Figure 1.4. Protein Production from Amino Acids.

Hence there seems to be a lot of 'molecular communication' going on between the parasite and their hosts. To explain this communications, some scientists have come up with a protein-based explanation. Before that we first need to understand how are proteins made and why they are important. To function normally a cell needs to run certain operations such as cell replication and self-repair after injury. Specific cells are tailored for certain functions. For instance, outer skin cells cover and protect the interior of our body from the environment while cells in the tongue have taste receptors that contribute to our sensation of taste. The biological minions that carry out these functions are proteins. Proteins are generated by a cellular factory called the ribosome. The ribosome produces proteins from a more basic component – amino acids. Based on the assembly of these amino acids, different proteins which perform various functions are developed. The instructions that dictate this assembly is obtained from the diriboxynucleic acid (DNA) which is in the nucleus of the cell. These proteins then go about the organism performing various jobs based on their functionality (see Figure 1.4). By producing proteins similar to their hosts, the parasite assimilates and cloaks itself within the host. Hidden this way, the parasites use these proteins to command the host's body in any way it pleases (Biron et al., 2005).

In Transit

Parasites often control the behavior of their hosts for their own benefit. The hairworm, parasitic wasp and the mermethid worm uses its hosts to secure and nurture their offspring. Parasites sometimes use other organisms for transport to get to their desired 'real host'. A good example of this type of behavior could be observed in flatworms. Flatworms are defined as trematodes (not nematodes). Among these worms is one Leucochloridium paradoxum which live in birds and uses snails for transport to get to birds. This flatworm lays its eggs in the birds which release it in its excrement. The excrement is then consumed by the amber snail. The eggs remain intact in the snail's digestive tract infecting it once it hatches. The juvenile flatworm (larvae) wanders around the amber snail's body infecting its internal organs. The larvae that reaches the snails head grows into its next developmental stage and begins replicating itself producing a 'brood sac' of sorts. During this process the larvae gradually invades the snail's eye stocks making it colorful. The eye stalks then swell and pulsate with colorful green and yellow bands. Then the parasite controls the snail's behavior such that it not only leaves the safety of the dark but starts climbing to tree tops where it could become easy prey to birds. The exposure to light further stimulates the pulsating behavior of the brood sac at the eye stock of the snail – making it look more like a caterpillar and thus more attractive to birds. And of course at some point, along comes a bird and the snail becomes a meal. Once in the bird, its 'real host' the mermethid worm lives and procreates. The worms lay eggs which will be released when the bird poops – giving rise to a whole new generation of worms which repeats this vicious and disgusting cycle.

Of course one way to view this is; that the parasite is transporting itself to its desired host while jumping through intermediate host(s) along the way. Or perhaps this was the case initially but in time the parasites started to evolve in such a way that it has to live its various stages of its life in various hosts. A good example of a parasite which inhabits multiple hosts is a type of flatworm called blood flukes. Flatworms cause a disease in humans called snail fever or Schistosomiasis affecting 200 million people worldwide. These types of flatworms usually live in fresh water and require two hosts in their life cycle – snails and humans (or other mammals). The eggs of the blood fluke are released by infected humans via urine or feces. With the capability of surviving a whole week on dry land, the eggs then end up in fresh water. The blood fluke larva then creeps into the snail via its foot. Once in the snail it progresses to its next stage of larva life (called sporocyst). It then asexually multiplies in the snail for months (it reproduces without a mate). It then moves out from the snail and waits patiently in the water for its human hosts to arrive. The moment they sense humans in the water, they quickly swim and attach themselves to the skin using suckers. Once the blood fluke larvae find a good location like a hair follicle, it quickly penetrates into the skin using enzymes. It then gets to specific blood veins via the blood stream. Once localized around specific human organs, the blood fluke parasites on blood cells – sucking up nutrients. The blood flukes continue by releasing their eggs which travel to the human intestines and bladder. The eggs are then released in fecal matter and urine and so the cycle continues. Further research has given us some idea on how parasites like the blood fluke can remain in the human body and evade attacks from the immune system.

The immune system of humans (as well as other mammals) immediately responds if it detects something foreign in its body. Parasites effectively evade these attacks via ‘camouflage’. Interestingly once the parasite gets into the mammalian host, it starts absorbing the host molecules into the surface of its body. This way the body fails to recognize the existence of the parasite since it camouflages as though it’s part of the host’s body. The second mechanism is called ‘molecular mimicry’. Biological functions in organisms are carried out via proteins (e.g., cell replication, repair, etc.). The parasites, therefore, produce various proteins which mimic the host proteins to carry out parasitic functions. Some of these proteins go even further and tamper with the host’s immune response facilitating the parasite’s survival (Tallima et al., 2015; de Oliveira Fraga et al., 2010).

THE POWER OF EXTENDED PHENOTYPES

Some parasitic behavior has been associated with the concept of ‘extended phenotype’. The idea was initially introduced by the evolutionary biologist, Richard Dawkins. Firstly a phenotype is an organism’s trait, behavior or characteristic which can be observed – as opposed to genotype which is the observable information in its genes. Therefore, the phenotype of an organism is to some degree influenced by the genotype. Other phenotypes could also be acquired from the environment. According to the extended phenotype, genes do not only influence the organism but its grasp extends further reaching its environment (or other organisms) (Dawkins, 1999). This of course can be seen frequently in many organisms. For instance, the human gut is a home for a wide range of non-harmful bacteria. These bacteria are critical for our digestion process. Therefore, the phenotype of the bacterial genes does not only ensure its own survival but extends to the digestion of food – which in effect contributes to our continuous survival.

A good example of extended phenotype in action is the cases of host manipulation by parasites discussed above. The parasite's genetic makeup thus influences itself as well as the host's behavioral responses such as molting, climbing to tree tops and triggering hormonal responses. In light of this, it is no surprise that there exists an effective molecular communication between host and parasite. The genetic material in the parasite has an extended phenotype allowing it to easily take control of the host. Using a computing analogue; both parasite and host have been installed with the same operating system. Therefore, swapping computer programs among both organisms would not be a problem and could be carried out easily.

Besides other organisms and the environment, the extended phenotypes of parasites have been seen to manipulate colonies of other organisms influencing the entire ecosystem. The powerful grasp of such extended phenotypes could be observed in a particular species of tapeworm called the *A. Brevis* (Hurd, H., 2015). This tapeworm infects two hosts; a certain ant and the spotted woodpecker. Like most tapeworms it lives and grows in the woodpecker's intestines. The tapeworm then lays eggs which are released by woodpecker in its feces. The ants gather these eggs and feed them to their young infecting them. The larval tapeworms hatch from the eggs and develop in the ants. The woodpecker completes the parasite's lifecycle by preying on the infected ants. The strange thing is that when the ants get infected, they become socially dysfunctional and start begging other ants in the colony for food. These ants could be easily identified since they change appearance by becoming golden yellow (in contrast to the brown color of other ants). Colors aside, even their odor starts to change. The thing about ants is that they recognize each other from the same colony by smell. This way they can quickly identify intruder ants invading their colony. The infected ants start developing various odors confusing the ant colony in recognizing its own colony's members. Colony members then start becoming less aggressive to intruders of different colonies – aiding in the spreading of the tapeworm's parasitic ventures.

This way the entire colony's security becomes compromised. In infected colonies, more ant larvae survive although they suffer from a low survival rate. This poses an additional burden on the colony since they would have feed and groom these inefficient members. The low survival rate is most probably due to their increased vulnerability to the spotted woodpecker. Strangely enough, to increase the likelihood of its transfer to its next host, the tapeworm increases the lifespan of the host ant. The extended phenotypes of parasites not only manipulate the host organism and its environment but could rather influence whole colonies. The powerful grasp of parasites on the ecosystem should not be underestimated.

FRAGILE BIODIVERSITY

In the past year a devastating fungal parasite called *Batrachochytrium dendrobatidis* (Bd in short) also known as amphibian chytrid has wiped out thirty percent of the world's amphibian population. The parasite has been known to cause a hundred percent mortality rate in some amphibian species. When infected with high amounts of Bd fungus, a fatal disease called *Chytridiomyosis* starts to manifest. Bd has been known to infect one third of amphibian species worldwide causing many species in Australia and South America to go extinct (Skerratt et al., 2007). Amphibians play an important role in our ecosystem – ensuring a critical balance that indirectly affects our own continuous survival. As of now Bd is most probably responsible for most disease-caused losses of biodiversity in the history of our planet (Whittaker and Vredenburg, 2011). The fungus Bd spreads by the releasing zoospores. Spores are seed-like units which act as reproductive agents for many types of plants and fungi. The mechanism of spore dispersal is an asexual reproduction for some of these life forms. Much like regular spores, zoospores are like small seeds with an additional component called a 'flagellum' which it uses to move or swim about. These zoospores swim through water and other moist environments. Zoospores are very resilient and can survive in wet environments for several months – though they will perish easily in a few hours if exposed to heat and dryness. Zoospores spread primarily by water or contact with amphibian skin. When Bd infects an amphibian, it damages the skin by off-balancing its electrolytes blocking key nutrients from passing through the skin. Bd infection then causes it to become lethargic and anorexic – eventually causing the animal to have a fatal cardiac arrest. Most amphibians that live or reproduce in streams and waterways become easily infected.

The resilience of the Bd fungus to live independently without hosts makes it more threatening. The effectiveness of the spread of Bd infections has been attributed to a few factors though its real cause remains unknown to this day. Among them is lowered resistance in amphibians to combat the fungus (Berger et al., 1999). These lowered resistances may have been due to changes in the environment e.g., habitat loss from logging and other human activities as well as global warming (Pounds et al., 2006). Another success of its spread is due to intermediate hosts – as discussed in the previous sections. The recent re-emergence and spread of the Ebola virus is similarly caused by intermediate hosts (fruit bats). Transferring the Ebola virus to humans the fruit bats themselves are not affected by the virus. Similarly the Bd fungus has been so far known to be able to survive in other non-amphibian organisms such as crayfish and waterfowls – making them very suitable intermediate hosts to get to the amphibians (McMahon et al., 2013).

Besides intermediate hosts, humans play an important role in spread of the Bd fungus. We often take amphibians and fish and release them in other parts of the world for reasons such as research, pets and even food. Along with the animals, the zoospores travel making its way into various habitats around the world, leaving a trail of dead amphibians behind. One actively traded amphibian is the bullfrog. Five million live bullfrogs are shipped annually into the U.S. to be sold as food. Bullfrogs are farmed as a food source in South America as well other parts of the world. Biologists at the University of Michigan studied the effects of bullfrog trade in the spread of Bd between the forests and farms of U.S., Brazil and Japan (Schloegel et al., 2012). The scientists found that 41 percent of the frogs sold in seven states in the U.S. were infected with Bd. Luckily Bd is only harmful to amphibians and not humans – making it safe to consume. These bullfrogs were said to be imported alive from Taiwan, Ecuador and Brazil (since Brazil's Atlantic forest is one of the most amphibian-rich places in the planet). They found that the Bd strain in Brazil was also found in Japan. The interesting thing about the North American bullfrog is its resistance to the Bd fungus although it does act as a carrier (or an intermediate host). Another team of biologists from the University of Colorado stated that the emergence of more virulent strains of Bd was most likely due to human trading (Erickson, 2012).

This is of course if we are talking about legal animal transport – what about illegal animal smuggling? Today animal smuggling provides for the black market trade for exotic animals. These animals are kept as exotic pets or killed; where various animal parts are harvested for local cuisines as well as traditional medicine. Such illegal smuggling provides an excellent mode of international transport for parasites. The sudden introduction of foreign parasites shocks the immune defenses of native animals. These animals then readily become hosts with minimal resistance (Daszak et al., 1999). Such novel interactions between hosts and parasites not only strengthen the parasites infectious diversity but also give rise to the emergence of whole new diseases.

Once transmitted into new environments via foreign hosts, the parasites have a field day altering the ecosystem of the foreign habitat – further threatening the biodiversity of the various native fauna. One way this can happen is when the parasite drives the host population towards extinction. Another way is via indirect effect where the parasite like Bd kills the frogs which prey on say flies. These flies then become abundant and start consuming more of some plant species. Similar interactions happen, causing ripple effects in the ecosystem modifying and restructuring it. Besides wiping out species, parasites like Bd also alter the ecosystem in a more subtle manner. Before killing its host Bd alters the host behavior e.g., inducing water-seeking behavior, reduced motion and basking. These sorts of behavior may increase the host's chances of becoming prey – driving up predation (Kilpatrick et al., 2010). Again, these types of interactions start subtly manipulating the different populations of the ecosystem.

The good thing is some amphibians like the North American bullfrog, the fire-bellied toad and certain lowland leopard frogs found in Arizona seem to be resistant to Bd infection. In 2011, a group of scientists from Cornell University studied these leopard frogs and found that resistances to the fungal had genetic origins (Black, 2011). Most animals on the planet including humans carry a segment of DNA called MHC (or major histocompatibility complex). The MHC identifies and performs sampling of the invading entity (in this case the Bd fungus). This sample is then carried to the immune system where it is recognized as a problem which needs to be resolved. Immediately the immune system starts responding to the fungal attack until the invader is neutralized. The frogs that survive the Bd attack thus carry the appropriate MHC genes triggering an immune response that counters Bd. The Cornell group found that the frogs that were most strongly exposed to the Bd fungal survived (Savage et al., 2011). This is due to what is called ‘selection’ in evolution lingo. It works like this – first due to strong Bd exposure most frogs are wiped out due to infection. However, a couple of them that contain the appropriate MHC genes that could initiate an immune response survive. Since the fungal exposure is intense, only the frogs that have the most effective MHC genes could reproduce and give rise to the next generation of frogs. These frogs may be more resistant to Bd since they acquired more strong MHC genes from both their parents.

The discovery of Bd-resistant MHC genes gives conservationists a new weapon to battle amphibian extinction. They started organizing recovery programs breeding Bd-resistant amphibians in captivity. These amphibians are then released back into the wild to breed with non-captive amphibians to propagate the Bd-resistant MHC genes. This leads to more Bd-resistant amphibians out in the wild, as stated by Associate Professor Bruce Waldman of Seoul National University, ‘We could selectively use those individuals with MHC molecules associated with resistance to the fungus for captive breeding programs. Susceptibility to the fungus depends on multiple factors, such as other components of the immune system, which need to be taken into account if we are to develop successful management strategies'. The catch-and-release recovery programs could be thought of as artificially aiding evolution to run its course at an increased pace with minimal species casualties.

Another strategy to curb the influence of Bd infections is by reducing host (amphibian) density in high susceptibility regions in order to prevent disease outbreak (Barlow, 1996; McCallum et al., 2001). High susceptibility regions here are places which are moist and warm suitable for the host-independent survival of the Bd fungus. This was done by collecting the amphibians and relocating them temporarily in another habitat or breeding them in captivity until it is safe again for exposure. This sort of culling is done during the initial stages of the outbreak or by predicting susceptible infection locations. Besides, the old fashion way of chemical treatment of the amphibians and their habitat may be helpful (Lilley et al., 1997). Using antibiotics or antifungal on the amphibians may directly combat the infection. Spraying their habitat with antifungal or releasing a certain dose of antifungal into the waterways may mitigate the fungal transmission and weaken its propagation. Weakening the Bd fungus makes it more vulnerable to predation or perhaps destroyed by other competing parasites and pathogens. Such dynamical interactions may be manipulated if we have a good understanding of the ecological powers at play as well as the implication of our mitigation instruments. Although chemical treatment may be effective, they may also alter the ecosystem by causing fatalities in other animal and plant species. Hence before chemical treatment, rigorous and thorough analyses relating to its potential effects need to be ascertained. If we are not careful, chemical treatments could create more damage than the fungus itself or even worst still – make the fungus more resistant and, therefore, more virulent affecting not only amphibians but many other types of organisms in the ecosystem.

One potentially effective methodology is vaccination following the path previously laid out by Pasteur and company. Similar to humans, vaccination may fortify amphibians against Bd by strengthening their immune system (f et al., 2011). As mentioned in the initial sections of this chapter, vaccination has worked very well for humans and other animals when it comes to battling serious epidemics like cholera and small pox. The virulence level of Bd has been known to differ among the various strains (Australian strain, South American strain, etc). Evidence has been found that the strains highly differ in terms of their genetic makeup. To develop a vaccine, we would need to identify factors that contribute to the weakening of the Bd strain. One thing we know for sure is that the degree of moisture and warmth in the environment affects its survival. If we try to correlate things like environmental temperature, moisture and nutrient availability to the level of virulence of Bd, then proceeding with developing a vaccine wouldn't be much trouble. The key here is identifying a weak version of the fungus such that when introduced into the amphibian it induces a strong response from its immune system fortifying it from future infections. Once we manage to alert the immune system, the game is up and the rest is done by immune system itself – recall the story of Jenner and the milkmaid in this chapter's first section. Now we should keep in mind that in some ways we have indeed come a long way from the time of Pasteur – at least in our knowledge of genetics. Genetic engineering is another avenue that could be explored for vaccine production. The idea here is to perform genetic modification to the existing Bd fungus and reproduce it in the lab. Such modifications would be done with the aim of generating a low virulence version of the fungus. If this is achieved, then similar like the catch-and-release recovery program mentioned above – instead of releasing resilient frogs, the weakened version of the Bd fungus is released. This Bd fungus would go on and naturally vaccinate the amphibians so that when the real highly virulent Bd arrives, immunity would have already been achieved. Besides releasing it into the ecosystem, individual amphibians could also be captured and vaccinated by scientists and conservationists. These individuals are then released into the wild to procreate whilst spreading the resilient gene to the next generation.

Ecological Engineering and Large Blues

First discovered in 1795, the Large Blue Butterfly (*Maculinea arion*) participates in a variety of complex interactions with other members of its ecosystem. Already considered a rare insect, the Large Blue mysteriously went extinct in the British Isles around 1979. Many organizations grew concerned and designed conservation programs directed towards the reintroduction of these butterflies in the British Isles using butterflies brought from Sweden (UK Butterflies, 2016). Although quiet successful, the conservation efforts really made a difference when scientists started comprehending lifecycles of Large Blues and using this knowledge to tap into their ecological interactions.

There are three key players in this ecological story; the Large Blue Butterfly, a species of tiny red ants (called *Myrmica sabuleti*) and the oregano plant. Large Blues actively search for Myrmica ants for the purpose of reproduction. First they lay their eggs on the buds of oregano plants. These eggs then hatch into Large Blue caterpillars. These caterpillars generate a scent that tricks the Myrmica ants into believing that they are ant larvae. The Myrmica ant then carries the caterpillar to their ant nests. In the ant nest the Large Blue caterpillar parasites on the Myrmica ants feeding on its larvae gradually destroying its colony. Scientists even found that these caterpillars even produce sounds that mimic the queen ant pretending to be elite members of the ant colony (Barbero et al., 2009). After ten months of parasiting, the caterpillars pupate and emerge as butterflies.

Now we look at the relationship between the Myrmica ant and the oregano plant. The Myrmica ant invades the oregano plant destroying its roots. To protect itself, the oregano plant wages a chemical warfare against the ants – secreting a chemical agent called carvacrol which kills the ants (Patricelli et al., 2015). It was found that the Myrmica ants respond by neutralizing the carvacrol effects by activating some of its genes. More carvacrol is released if the plant is under heavy attack from Myrmica ants. Besides being an ant repellant, carvacrol also produces a warm odor which is picked up by Large Blues. These Large Blues then proceed to the oregano plants (which are being attacked) to lay their eggs on the buds. The eggs hatch and the caterpillar makes its way to the ant nest and destroys the ant colony protecting the oregano plant. The oregano plant is thus seen to put up a distress call for the Large Blues when under attack by Myrmica ants.

Figure 1.5. Interaction between Large Blues, Myrmica ants and oregano plants. (Source: Patricelli et al., 2015)

Therefore, the oregano plant sacrifices some of its leaves to the caterpillar and some of its roots to the invading Myrmica ants. To hinder a lethal invasion it couples two types of strategies; chemical warfare and calling for reinforcement (Large Blues). These then attacks the ants while maintaining the population of the Large Blues by aiding its reproduction. To avoid complete extermination, the Myrmica ants fight back by neutralizing the chemical agent carvacrol. This delicate ecological balance shows how lives of the parasite (Large Blues), the oregano plant and the Myrmica ant are interwoven (Figure 1.5).

Conservationists discovered later that one of the reasons contributing the disappearance of Large Blues in the British Isles is the change in grazing habits of rabbits which also share the same habitat. These rabbits were reducing in population after contracting a fatal viral disease (*myxomatosis*). This left the grassland too tall and shady causing the ants which love heat to lose interest. Without the ants, the Large Blues could not reproduce decreasing its population. Although the reintroduction of Large Blues from Sweden has greatly improved its numbers in the British Isles, scientists have now known that its population is more significantly affected by two other interdependent members; the oregano plant and the Myrmica ant. Understanding the intricate interactions of parasites and hosts along with other members in the ecosystem effectively benefits conservation efforts.

SIMULATING THE ECOSYSTEM

Researchers have found that computer models are very effective in simulating the interplay of ecological forces and their impacts on the ecosystem. Such models are becoming indispensable when it comes to conservative efforts. An interesting example of this is the study on reintroduced animal populations. A popular strategy for animal conservation is the reintroduction of an endangered species/population. Keeping the Large Blues conservation in mind; the declining butterflies were reintroduced into the British Isles using stock from Sweden. This increases the breeding rate of the dying population recovering the population to its original numbers. For the reintroduction approach to work, the introduced alien population would have to be accepted as potential mates by the local population. This is where it gets a little tricky since no one knows how to obtain information on the 'acceptance rate' of the alien population. Besides, nothing is even known about what happens to this alien population once reintegrated.

Ecology and conservation researchers from the University of Florida used a computer simulation model to assist them in obtaining this information. The simulation model was built based on advanced circuit theory aimed at understanding animal movement. Using data from chimpanzee populations, the researchers modeled the mate choice and dispersal rate across multiple generations. The simulation showed that the inability for the alien animals to mate with the local population could result in habitat fragmentation – where the alien animals live and breed separately from the local population in their own habitat. The researchers stated that the physical toll taken on the alien population due to transport and its struggle to adapt to the local environment often makes them undesirable to the local mates. They also noted that it is possible that females in the local population to be less attracted to the alien animals due to their unfamiliarity. Familiarity is a key component for mating success among organisms. This is because familiarity allows the potential mate to be evaluated which in return produces fit offspring. The other side is that the local animals would prefer to mate with alien animals since this way they could avoid inbreeding which greatly affects the fitness of the offspring. The computer model developed greatly contributes to our understanding on the effectiveness of conservation strategies such as reintroduction and the parameters that we should look out for when developing conservation approaches.

It would be really nice to have a computer model that allows you to test different factors influencing ecosystems. For instance, to study the ecological effects of increasing the population of one species or reducing that of another. Or knowing if there are possible parasites lurking about. Something more interesting would be a simulation that gives you hints of possible hidden features of processes in the ecosystem. Even a model that provides us with roles of the organisms in the ecosystem would be a great help. The closest we have come to such a fantastical simulator is the Madingley model developed by a joint research between Microsoft Computational Science Lab and United Nations Environment Program World Conservation Monitoring Centre (Harfoot et al., 2014; Smith, 2014). The Madingley model simulates various aspects of ecosystems; organism migration, population growth and their specific roles in the food chain. It even maps the flow of energy and nutrients through the ecosystem over time.

As the scientists involved in the model development put it – 'Our results indicate that ecologists have gathered sufficient information to begin to build realistic, global, and mechanistic models of ecosystems, capable of predicting a diverse range of ecosystem properties and their response to human pressures. Ecosystems across the world are being rapidly degraded. This threatens their provision of natural goods and services, upon which all life depends. To be able to reduce – and one day reverse – this damage, we need to be able to predict the effects of human actions on ecosystems.' And this is exactly what the model primarily aims to do. The model is built on five basic processes which are fundamental to all ecosystems on our planet; metabolism, eating, dispersal, death and reproduction. When these principles are added with other, considerations such as body mass, organism type and the role of specific members in the ecosystem could be determined. The Madingley model is aimed much higher to someday simulate all of life on earth, calling it the Global Ecosystem Model (GEM).

As of now the Madingley model is capable of giving us information regarding impacts of change in environmental conditions (e.g., the looming threat of climate change), habitat loss from human activities such as harvesting. The model even gives some idea on trends in biodiversity. Something like the Madingley model is very helpful in predicting loss of biodiversity from parasites or disease. For instance the model would be able to tell what happens to the ecosystem if amphibians become extinct. This way we could really understand how ecosystems behave and the kind of repercussions we would expect in the face of species extinctions (see previous section: Fragile Biodiversity). The great thing about the Madingley model is that it's open-source and freely available online starting in 2014. The simulation is initiated by specifying things like; location size, terrain type (ocean, land or fresh water) and physical conditions (temperature, humidity, etc.). Then you introduce herbivores, carnivores and other organisms. Once completed the simulation runs and predicts the effects with time after multiple generations.

Still in its developmental stage the Madingley model does not provide highly detailed predictions. Nevertheless as more researchers improve the model's various formulations, it's only a matter of time until we obtain a very accurate model for global ecosystem simulations. The key element used for expediting scientific discoveries is the internet. Besides enabling discussions and information distribution, the sharing of tools is developing into a culture among scientists. Open-sourcing models and other scientific tools without charge allow scientists from different corners of the globe to share their ideas and contribute at a faster rate. Besides, the sheer diversity of ideas shared allows more research obstacles to be overcome. Assuming that the culture of open-sourcing science continues, the continual speedy improvement of simulators such as the Madingley model is assured. The Head of Microsoft Research on Computational Ecology and Environmental Science and one of the founders of the Madingley model, Drew Purves states, 'Our model is a first working version which will hopefully encourage other scientists to become involved in developing this, or analogous, global models of life.'

Gaining from Natural Experiments

Another useful approach that could be used for understanding ecological effects is the idea of 'natural experimentation' which is frequently utilized in economics research. Structures such as economic systems are so vast and complex that tampering with them causes large-scale repercussions and irreversible effects. Therefore, conducting experiments on such systems are next to impossible – with the exception of virtual experiments using computer simulations like the Madingley model. An interesting technique often employed in economics to study such systems is 'natural experimentation'. This approach uses just observation where the researcher has no control of the parameters influencing the experiment. The experiment is actually a naturally occurring phenomenon where the phenomena occurs in such a way that it could be used as an experiment. The experimental results are then employed to answer scientific questions. Simply put if we cannot design an experiment to test a hypothesis in the lab, then perhaps we ought to look out for naturally occurring phenomena that could be used as an experiment. Of course it would be very convenient if the natural experiment mimics closely the conditions dreamed up by the researcher. Economists use natural experiments as a testing ground when faced with costly experiments which are too complex scale wise or unethical (if it involves human subjects).

An interesting application of natural experiments is the study of baby birth weights with respect to the level of air pollution (Rich et al., 2015). Although some experiments have been conducted on analyzing the effects of pollution on baby weights during birth, the findings have been largely inconsistent due to several factors. For instance; a sudden decrease in air pollution is required so that the results of baby weights during high levels of pollution could be compared. Inducing circumstances where there is a sudden drop in air pollution in a huge city is not possible. Thanks to the Beijing Olympics in 2008, researchers had the opportunity to catch hold of just the right natural experiment. In preparation for the Olympics, the Chinese government put into place a series of environmental regulations to decrease the air pollution in Beijing. These policies include the temporary suspension of construction projects and closure of factories in addition to artificially stimulating rain via cloud seeding. These measures temporarily improved the air quality in the city during the time of the Olympic Games – the exact comparison the experiment required. Researchers then quickly used this situation as a natural experiment studying the effects on the weight of 83,672 babies born in Beijing during the games. The results were then compared with the weights of the babies born during the same period of 2007 and 2009 (during heavy pollution). The researchers discovered that the babies born during the low-pollution window of the 2008 Olympics weighed 23 grams more than those born during the periods of 2007 and 2009 (before and after the games). Thanks to the natural experiment, the studies successfully showed that air pollution does affect the birth weights of babies.

So that's how natural experiments work in economics. Although not widely explored in ecology, there are some instances where this approach has been effective. Some ecologists have considered using long-term natural experimentation as a strategy to understand the effects of vegetation change and biodiversity (Wardle et al., 2012). These studies were conducted to understand how an ecosystem declines over time from major disturbances – such as fires, flooding, climate change, insect outbreaks or impacts from human activity. And how this decline affects the change in vegetation and biodiversity. These scientists conducted their natural experimentation by observing the ecology of 30 islands in the Swedish boreal forest that form over a 5000 year period. These islands had various exposures to forest fires. This allowed them to study similar ecosystems with various levels of forest fires. Without natural experimentation, these studies would be otherwise impossible to conduct since we cannot go around starting forest fires for scientific purposes. Firstly, they cannot be contained and the damage on the ecosystem is irreversible. Thanks to natural experimentation their studies uncovered some of the critical factors influenced by the forest fires; which in effect drives the ecosystem into decline.

The study of disease vulnerability in Galapagos hawks successfully utilized natural experiments (Whiteman et al., 2006). This research was centered on figuring out how genetic diversity links with disease resistance in Galapagos hawks. Such a study could not be performed in a laboratory setting. For one, it is quite unethical to inbreed animals and see if they get sick easily. Second, the sheer number of these animals that would be needed for the experiment would be enormous. And this would require plenty of funding. Understanding disease vulnerability among island animal species is very crucial for conservation. This is because more than

75% of animal extinction including 90% of bird extinction is caused by disease spread in islands (Reid and Miller, 1989). Since the Galapagos Islands vary in size, a unique opportunity for natural experimentation presents itself. Small islands often encourage inbreeding among Galapagos hawks due to scant supply of potential mates. The situation is reversed for large islands where Galapagos hawks have the most diversity.

Parasites and hosts are constantly in an evolutionary struggle. With time the parasite evolves new traits making it more adaptable to the host. The host on the other hand develops new defense mechanisms to be used against the parasite. This continual struggle carries on – but to what end is hitherto not clear to science. If the host organism has good genetic diversity, then it can develop much richer adaptations. So when it gets invaded by a parasite, its defenses are better prepared to counter the parasite. Poor genetic diversity of hosts causes parasites to easily weaken the host's defenses. Researchers tested this theory using natural experiments on Galapagos hawks. These hawks are usually infected by two types of chewing louse parasites. Upon infection, the hawks fight off the parasites by triggering an immune response which is measured by taking their antibody count. The researchers found that Galapagos hawks on larger sized islands with richer genetic diversity have a stronger immune response resulting in greater defenses against the louse parasite compared to the inbred small island hawks.

Their findings highlight the impact of introducing foreign organisms (or humans) to small islands. Since the inbred small island fauna have weak defenses against parasites, even the introduction of fairly harmless foreign parasites are detrimental. Leaving these islands in isolation and avoiding travel may be recommended to avoid further loss in biodiversity. Or perhaps we could try some ecological engineering to gradually increase the biodiversity of small islands making the fauna more resilient to parasites. One way is by introducing genetic diversity gradually by transferring species from nearby islands from time to time. Perhaps a good ecological simulation could give some idea on:

1. Degree of diversity.
2. Amount of population needing to be introduced.
3. Systemic balance of the ecology as a result of the species introduction.

Recently a group of American biologists employed natural experiments to investigate the impact of loss in predatory birds on their insect prey (spiders) (Rogers et al., 2012). Similar to the Galapagos hawks; lab-based experiments are too difficult to perform due to the scale of the study. Here the research encompassed four Western Pacific islands; Guam, Rota, Saipan and Tinian. In the island of Guam, a type of brown tree snake was introduced in the mid-1940s. Insect-eating birds are the natural prey for the brown tree snake. This introduction caused a severe population decline among these Guam island birds. On the other three Western Pacific islands; Rota, Saipan and Tinian, the brown tree snake was not introduced. This situation gave a unique opportunity for studying the impact of insect-eating birds on the ecosystem over the time frame of about 25 years. Conventional experiments running through such lengthy time scales are definitely not feasible, not to mention the impossibly high funding required. Using this situation as a natural experiment, the biologists found various interesting results. They discovered that spiders, the bird's natural prey were thriving on Guam Island as compared to the other three islands (where birds were present). This is because the introduced brown snake preyed on most of the insect-eating birds on Guam exterminating the spider's natural predator. Compared to smaller lab-scale studies, the researchers found that the impact of bird loss on spider populations were far greater than anticipated.

Besides predator-prey interactions of birds and spiders, other factors may explain the sudden explosion in the spider population. For instance, without birds the spiders' webs are not constantly torn away by birds flying through them. And birds often use spider webs for constructing their nests. In the absence of birds, spiders spend less energy re-building these webs; allowing them to focus in reproduction. The researchers also found that under no threat, the spiders build much larger webs, increasing their chances of capturing prey. Such an advantage increases the security of the spider's food supply letting them procreate with increasing frequency.

In Brief

Starting with Avicenna, Rhaze and Hipppocrates, the study of parasites was strengthened in the very capable hands of Edward Jenner, Louise Pasteur and Robert Koch. Beginning with parasite worms, thanks to the invention of the microscope we soon started becoming aware of other disease-causing parasitic microbes. The life-cycles of parasites are highly complex and they often employ various strange host manipulation strategies. Some of their strategies involve; using multiple hosts, cloaking against the host's defenses and biochemical mimicry. Parasites such as Bd have significantly devastated the biodiversity of amphibians on the planet. Ecological engineering is effective for conservation. To aid such efforts, computer programs that simulate the ecosystem such as the Madingley model prove to be very helpful. Such simulators help in giving some idea on how ecosystems react to various pressures – such as human activity and climate change. Similar to simulators, natural experiments are very effective in providing essential information about the ecology.

References

Ashford, R.W. and W. Crewe. 1998. *The parasites of Homo sapiens*. Liverpool School of Tropical Medicine, Liverpool, United Kingdom.

Barbero, F., Thomas, J.A., Bonelli, S., Balletto, E. and Schönrogge, K., 2009. Queen ants make distinctive sounds that are mimicked by a butterfly social parasite. *Science,* 323(5915), pp.782-785.

Barlow, N.D., 1996.The ecology of wildlife disease control: Simple models revisited. *J Appl Ecol,* 33:303-314.

Berger, L., Speare, R. and Hyatt, A., 1999. Chytrid fungi and amphibian declines: Overview, implications and future directions, *Declines and Disappearance of Australian Frogs,* pp 22 – 33.

Biron, D.G., Marché, L., Ponton, F., Loxdale, H.D., Galéotti, N., Renault, L., Joly, C. and Thomas, F., 2005. *Behavioural manipulation in a grasshopper harbouring hairworm: a proteomics approach. Proceedings of the Royal Society of London B: Biological Sciences*, 272(1577), pp.2117-2126.

Black, R., 2011. A Frog killer immune genes revealed, Science & Environment, *BBC News*: http://www.bbc.com/news/science-environment-15060980.

Blevins, S.M. and Bronze, M.S., 2010. Robert Koch and the 'golden age' of bacteriology, *International Journal of Infectious Diseases,* Vol 14, pp 744–751.

Bryan, C.P. 1930. *The Papyrus Ebers (translated from the German).* Geoffrey Bles, London, United Kingdom.

Buckman, 2002. *Human Wildlife: The Life That Lives on Us,* Johns Hopkins University Press.

Cambau, E. and Drancourt, M., 2014. Steps towards the discovery of Mycobacterium tuberculosis by Robert Koch, 1882. *Clinical Microbiology and Infection,* 20(3), pp.196-201.

Cox, F.E., 2002. History of human parasitology. *Clinical microbiology reviews,* 15(4), pp.595-612.

Dawkins, R., 1999. *The extended phenotype: the long reach of the gene.* Oxford Paperbacks.

Daszak, P., Berger, L., Cunningham, A.A., Hyatt, A.D., Green, D.E. and Speare, R., 1999. Emerging infectious diseases and amphibian population declines. *Emerging infectious diseases,* 5(6), p.735.

de Oliveira Fraga, L.A., Lamb, E.W., Moreno, E.C., Chatterjee, M., Dvořák, J., Delcroix, M., Sajid, M., Caffrey, C.R. and Davies, S.J., 2010. Rapid induction of IgE responses to a worm cysteine protease during murine pre-patent schistosome infection. *BMC immunology,* 11(1), p.56.

Dexiang, W., Wenyuan, Y., Shenqi, H., Yunfang, L., Tiancheng, S., Jiahua, M., Wenxiu, H. and Nianfeng, X., 1981. Parasitological investigation on the ancient corpse of the Western Han Dynasty unearthed from tomb no. 168 on Phoenix Hill in Jiangling County. *Acta Academiae Medicinae Wuhan,* 1(2), pp.16-23.

Erickson, J., 2012. Global bullfrog trade spreads deadly amphibian fungus worldwide, Michigan News, University of Michigan: http://ns.umich.edu/new/releases/20667-global-bullfrog-trade-spreads-deadly-amphibian-fungus-worldwide.

Ferreira, L.F., De Araujo, A.J.G. and Confalonieri, U.E., 1983. The finding of helminth eggs in a Brazilian mummy. *Transactions of the Royal Society of tropical Medicine and Hygiene,* 77(1), pp.65-67.

Glud, R.N., Wenzhöfer, F., Middelboe, M., Oguri, K., Turnewitsch, R., Canfield, D.E. and Kitazato, H., 2013. High rates of microbial carbon turnover in sediments in the deepest oceanic trench on Earth. *Nature Geoscience,* 6(4), pp.284-288.

Harfoot, M.B., Newbold, T., Tittensor, D.P., Emmott, S., Hutton, J., Lyutsarev, V., Smith, M.J., Scharlemann, J.P. and Purves, D.W., 2014. Emergent global patterns of ecosystem structure and function from a mechanistic general ecosystem model. *PLoS Biology,* 12(4), p.e1001841.

Hoeppli, R. 1959. *Parasites and parasitic infections in early science and medicine.* University of Malaya Press, Singapore, Singapore.

Horne, P.D. 1985. A review of the evidence for human endoparasitism in the pre-Columbian New World through the study of coprolites. *J. Arch. Sci.* 12:299–310.

Hurd, H., 2015. *A tapeworm's extended phenotype reaches further than its host,* BugBitten, BioMed Central.

Lilley J.H., Inglis, V., 1997. Comparative effects of various antibiotics, fungicides and disinfectants on Aphanomyces invaderis and other saprolegniaceous fungi. *Aquaculture Research*, 28:461-469.

Kilpatrick, A.M., Briggs, C.J. and Daszak, P., 2010. The ecology and impact of chytridiomycosis: an emerging disease of amphibians. *Trends in Ecology & Evolution,* 25(2), pp.109-118.

Koch R, 1877. Verfahren zur Untersuchung, zum Conservieren und Photographiren der Bakterien. [Method of examination, preservation and photography of the bacteria.] *Beitra¨ge zur Biologie der Pflanzen,* Vol 2, pp 399–434.

Koino, S. 1922. Experimental infection of the human body with ascarides. *Japan Medical World*, Vol 15:317-320.

McCallum, H., Barlow, N. and Hone J., 2001. How should pathogen transmission be modeled? *Trends in Ecology & Evolution*, 16:295-300.

McMahon, T.A., Brannelly, L.A., Chatfield, M.W., Johnson, P.T., Joseph, M.B., McKenzie, V.J., Richards-Zawacki, C.L., Venesky, M.D. and Rohr, J.R., 2013. Chytrid fungus Batrachochytrium dendrobatidis has nonamphibian hosts and releases chemicals that cause pathology in the absence of infection. *Proceedings of the National Academy of Sciences*, 110(1), pp.210-215.

Patricelli, D., Barbero, F., Occhipinti, A., Bertea, C.M., Bonelli, S., Casacci, L.P., Zebelo, S.A., Crocoll, C., Gershenzon, J., Maffei, M.E. and Thomas, J.A., 2015, July. Plant defences against ants provide a pathway to social parasitism in butterflies. In *Proc. R. Soc. B* (Vol. 282, No. 1811, p. 20151111). The Royal Society.

Poulin, R., 1998. *Evolutionary Ecology of Parasites: From Individuals to Communities.* Chapman & Hall, London.

Pounds, J.A., Bustamante, M.R., Coloma, L.A., Consuegra, J.A., Fogden, M.P., Foster, P.N., La Marca, E., Masters, K.L., Merino-Viteri, A., Puschendorf, R. and Ron, S.R., 2006. Widespread amphibian extinctions from epidemic disease driven by global warming. *Nature,* 439(7073), pp.161-167.

Reid, W.V. & Miller, K.R. 1989 *Keeping options alive: the scientific basis for conserving biodiversity.* Washington, DC: World Resources Institute Reid, W.V. & Miller, K.R. 1989

Rich, D.Q., Liu, K., Zhang, J., Thurston, S.W., Stevens, T.P., Pan, Y., Kane, C., Weinberger, B., Ohman-Strickland, P., Woodruff, T.J. and Duan, X., 2015. Differences in Birth Weight Associated with the 2008 Beijing Olympics Air Pollution Reduction: Results from a Natural

Experiment. *Environmental Health Perspectives (Online),* 123(9), p.880.

Rogers, H., Hille Ris Lambers, J., Miller, R. and Tewksbury, J.J., (2012), 'Natural experiment' Demonstrates Top-Down Control of Spiders by Birds on a Landscape. Level. *PLoS ONE* 7(9): e43446. doi:10.1371/journal.pone.0043446.

Ross, A.D., 2015. Zombie spider builds a stronger web for the parasitic wasp that's sucking its blood, *The Verge*, http://www.theverge.com/2015/8/6/9103979/parasitic-wasp-spiders-zombie-stronger-web.

Savage, A.E. and Zamudio, K.R., 2011. MHC genotypes associate with resistance to a frog-killing fungus. *Proceedings of the National Academy of Sciences*, 108(40), pp.16705-16710.

Schloegel, L.M., Toledo, L.F., Longcore, J.E., Greenspan, S.E., Vieira, C.A., Lee, M., Zhao, S., Wangen, C., FERREIRA, C., Hipolito, M. and Davies, A.J., 2012. Novel, panzootic and hybrid genotypes of amphibian chytridiomycosis associated with the bullfrog trade. *Molecular Ecology,* 21(21), pp.5162-5177.

Skerratt, L., Berger, L., Waldman, B., McFadden, M. and Hunter, D., (2016), *Five-year study unlocks global frog death mystery*, James Cook University, Australia: https://www.jcu.edu.au/news/releases/news-archive2/news-and-media11111.

Skerratt, L.F., Berger, L., Speare, R., Cashins, S., McDonald, K.R., Phillott, A.D., Hines, H.B. and Kenyon, N., 2007. Spread of chytridiomycosis has caused the rapid global decline and extinction of frogs. *EcoHealth,* 4(2), pp.125-134.

Smith, J., 2014. *Life boils down to five 'rules' ... or so says the Madingley Model, The Conversation.*

Takasuka, K., Yasui, T., Ishigami, T., Nakata, K., Matsumoto, R., Ikeda, K. and Maeto, K., 2015. Host manipulation by an ichneumonid spider ectoparasitoid that takes advantage of preprogrammed web-building behaviour for its cocoon protection. *Journal of Experimental Biology,* 218(15), pp.2326-2332.

Tallima, H., Dalton, J.P. and El Ridi, R., 2015. Induction of protective immune responses against Schistosomiasis haematobium in hamsters and mice using cysteine peptidase-based vaccine. *Frontiers in immunology,* 6.

Treonis, A.M. and Wall, D.H., 2005. Soil nematodes and desiccation survival in the extreme arid environment of the Antarctic Dry Valleys. *Integrative and Comparative Biology,* Vol 45(5), pp.741-750.

Vance, S.A., 1996. Morphological and behavioural sex reversal in mermithid-infected mayflies. *Proceedings of the Royal Society of London B: Biological Sciences*, 263(1372), pp.907-912.

Vasudev, D. and Fletcher Jr, R.J., 2016. Mate choice interacts with movement limitations to influence effective dispersal. *Ecological Modelling,* 327, pp.65-73.

Vinodbhai, P.N., 2013. Evolutionary Perspective of Human Papilloma Virus Infection in Humans. *Journal of Antivirals & Antiretrovirals,* 2013.

Wardle, D.A., Jonsson, M., Bansal, S., Bardgett, R.D., Gundale, M.J. and Metcalfe, D.B., 2012. *Linking vegetation change, carbon sequestration and biodiversity: insights from island ecosystems in a long-term natural exp.*

Whitman, W.B., Coleman, D.C. and Wiebe, W.J., 1998. Prokaryotes: the unseen majority. *Proceedings of the National Academy of Sciences*, 95(12), pp.6578-6583.

Whiteman, N.K., Matson, K.D., Bollmer, J.L. and Parker, P.G., 2006. Disease ecology in the Galapagos Hawk (Buteo galapagoensis): host genetic diversity, parasite load and natural antibodies, *Proceedings of the Royal Society B,* 273, 797–804.

Woodhams, D.C., Bosch, J., Briggs, C.J., Cashins, S., Davis, L.R., Lauer, A., Muths, E., Puschendorf, R., Schmidt, B.R., Sheafor, B. and Voyles, J., 2011. Mitigating amphibian disease: strategies to maintain wild populations and control chytridiomycosis. *Frontiers in Zoology,* 8(1), p.1.

Zolfagharifard, E., 2014. The computer that plays GOD: Scientists design simulator that predicts the fate of all life on Earth, *MailOnline*;

http://www.dailymail.co.uk/sciencetech/article-2611130/The-computer-plays-GOD-Scientists-design-simulator-predicts-fate-life-Earth.html.

Web Sources

Boese, Alex (11 August 2011). "Grassi Eats Worms." *Mad Science Museum.* Retrieved 25 October 2014 – < http://www.madsciencemuseum.com/msm/pl/grassi_ eats_worms>

History of Discovery, 2006 – < http://web.stanford.edu/group/parasites/ParaSites2006/Dracunculiasis/history.htm> UK Butterflies, 2016 – http://www.ukbutterflies.co.uk/species.php?species=arion

Louis Pasteur, 2016 – http://www.historylearningsite.co.uk/a-history-of-medicine/louis-pasteur/

Rabies, 2016 – <http://www.cdc.gov/rabies/>

Robert Koch, 2016 – <http://www.historylearningsite.co.uk/a-history-of-medicine/robert-koch/>

Whittaker, K. and Vredenburg, V., 2011. *An Overview of Chytridiomycosis*, http://www.amphibiaweb.org/chytrid/chytridiomycosis.html.

Zimmer, C., 2013. The Parasite's Fountain of Youth, The Loom, *National Geographic* – http://phenomena.nationalgeographic.com/2013/02/25/the-parasites-fountain-of-youth/

Chapter 2

Antibiotics, Vaccines and Epidemics

"Nothing is so painful to the human mind as a great and sudden change."

— Mary Shelley, Frankenstein

The Journey of Antibiotics

Antibiotics was conceived and applied extensively by medical practitioners since the early twentieth century. Earliest applications of antibiotics could be traced back to Nubia or modern-day Sudan. Residues of tetracycline were found on human skeletal remains buried there during the period of 350 -550 A.D (Bassett et al., 1980; Nelson et al., 2010). Tetracycline is a type of broad-spectrum antibiotic used for tackling a wide range of bacterial infections. And the drug is on the World Health Organization's (WHO) List of Essential Medicines used for skin infections (e.g., moderate to chronic acne and rosacea) and infections of the urinary tract, respiratory tract and intestines. It was found that the distribution of tetracycline in the bones of these Nubians could only be explained if tetracycline-based foods were consumed at that time. Besides Nubians, traces of tetracycline were also present in the remains of people living during the Roman period in the Dakleh Oasis, Egypt. Similar to Nubians, the residues in the bones indicated that those Egyptians had included tetracycline in their diet (Cook et al., 1989). In both cases (Nubians and Egyptians) the evidence shows a low rate of infectious diseases encountered by these communities. And there were no proof of past infections found on the uncovered remains. Consuming antibiotics, these communities managed to stay clear of infections compared to other societies at that time. Another interesting evidence of the employment of antibiotics is in Traditional Chinese Medicine (Cui and Su, 2009).

Qinghaosu or artemisinin has been used by Chinese herbalists and doctors for thousands of years to remedy many parasitic illnesses. Artemisinin could be found in Artermisia group of plants commonly known as 'wormwood', named for its peculiar characteristic for driving away worms (depicted in Figure 2.1). In Traditional Chinese Medicine, the Artemisia plant was used during the Qing Dynasty to treat what is termed summer-heat syndromes; since the ailments occur only during summer time (Wen and Seifert, 2000). These ailments are caused by parasites and commonly transmitted via contaminated water and food. The traditional remedy for alleviating malaria symptoms was first written in Ge Hong's 'A Handbook of Prescriptions for Emergencies' in 284 – 346 CE (Youyou, 2011). Malaria in humans is caused by protozoa (a type of microbes called *plasmodium falciparum*) and is transmitted by some species of mosquito. In the seventies, artemisinin was successfully extracted and its effectiveness against malaria was demonstrated in the eighties. Today artemisinin combination therapy is considered standard anti-malarial treatment. Tu Youyou, a Chinese scientist shared the 2015 Novel prize in Medicine for her work on the discovery and development of artemisinin. Today the WHO requires that artemisinin is always used in combination with other drugs for treating Malaria. Some ancient communities have been effectively administering some forms of antibiotics before the dawn of modern-day medical science. Due to this long-term exposure, it is highly probable that antibiotics resistances have emerged among humans at a genetic-level.

Figure 2.1. An illustration of Artemisia Annua plant during the Ming Dynasty (1591 CE) (left) and a photograph of Artemisia Annua in the field (right) (Youyou, 2011).

The antibiotic era began with the discoveries of penicillin by Alexander Fleming and Salvarsan by Paul Ehrlich. By observing synthetic dyes used for coloring microbes in the later eighteen hundreds, Paul Ehrlich was inspired to create a compound that targets specific microbes and not others. This way the compound targets only parasites and leave other microbes which are essential to the body unaffected. Ehrlich then invented the 'systematic screening' methodology which is still used today. Systematic screening or compound screening is the process of identifying a drug (known as the 'hit compound') which has a desired chemical or biological activity. This hit compound is identified from a library of candidate compounds. Today there are many variations of systematic screening approaches (Hughes et al., 2011). Using systematic screening Elrich together with two other scientists discovered an effective drug called Salvarsan – for effective use against the then endemic syphilis. Syphilis is caused by a type of bacteria (*Treponema palladium*) and is usually treated at the time using inorganic mercury salts. This treatment is extremely ineffective and causes severe side effects to patients. Salvarsan was the

606th compound developed and tested against the disease. It showed immediate success against syphilis-infected rabbits. In limited human trials, Salvarsan quickly proved to be highly effective against the sexually transmitted disease. The success of Salvarsan prompted the development of the soluble and less toxic drug, Neosalvarsan which became standard treatment for syphilis until it was replaced by penicillin (Mahoney, 1943).

Systematic screening gave birth to the discovery of a variety of drugs. Among them is a class of drugs called sulfa drugs. Sulfa drugs are used to treat cough and allergies. Besides, it also functions as an antibiotic, antifungal and antimalarial. Among the earliest forms of sulfa drugs is a compound known as Protonsil. Synthesized by chemists Josef Klarer and Fritz Mietzch, Protonsil is among the oldest antibiotics available in the market causing among the first cases of widely spread antibiotic resistance – sulfa drug resistance (Enne et al., 2004).

In the early nineteen hundreds, Alexander Fleming discovered penicillin. Fleming extracted penicillin from the activity of a species of fungus called *Penicillin notatum*. During the First World War, Fleming was Captain in the Royal Army Medical Corps working in hospitals at the western fronts of France. There he observed that many soldiers died from sepsis resulting from infected wounds. He noticed that antiseptics caused more harm to the patient's immunological defenses as compared to the killing of foreign bacteria. In an article he submitted to the medical journal, *Lancet*, Fleming described how antiseptics were killing more soldiers than the war itself. After the war, a serendipitous contamination of one of his bacterial cultures led to the discovery of penicillin (which he called 'mould juice' at the time). Twelve years after his discovery, Fleming persistently tried to attract the interest of chemists to solve the problem of purification of the substance. He did this by providing the penicillium fungus strain to any researcher requesting it. In 1940, an Oxford research group published a paper describing the purification of penicillin sufficient for clinical testing (Chain et al., 2005). Their methodology was employed to mass produce penicillin in 1945. Upon the discovery of penicillin, Fleming warned of the potential resistance to the drug.

The golden age of drug discovery could be attributed to the period of 1950s to 1970s where novel antibiotics were synthesized. Ever since then, resistance to antibiotics has been tackled by administering modified versions of the existing drugs (Aminov, 2010). Another version of antibiotic resistance is multi-drug resistance. This sort of resistance occurs when the patient has an infection which is resistant to an array of drugs. Approximately 25,000 patients die from infection of multi-drug resistant bacteria annually in the European Union alone (ECDC/EMEA, 2009).

How Antibiotic Resistance Works

Antibiotic resistance is a phenomenon that occurs in persons where the antibiotic losses its ability to effectively kill or control bacterial growth. Antibiotics and similar drugs have been used to treat patients with infectious disease for the last seventy years. In the United States, a minimal of 2 million people are infected with antibiotic resistant bacteria (Drug Resistanc – CDC, 2016). In such cases, the bacteria become resistant to the presence of antibiotic resulting in continual proliferation in the body. The mechanism of how this happens could be very elegantly explained using concepts from the theory of evolution (Alliance for the Prudent use of Antibiotics (APUA), 2014). Here it goes; when an antibiotic is introduced in the body, there would be some bacteria which would have higher chances of surviving as compared to others (which get killed by the treatment). This makes the surviving population of bacteria resistant to the administered antibiotic. So when all susceptible bacteria are killed by the antibiotic, the remaining fitter populations develop resistant genes.

Due to selective pressure, these resistant-bacteria continue to thrive producing the next generation of resistant-bacteria. They continue to multiply in the presence of the antibiotic (See Figure 2.2). Antibiotic resistance occurs when the administered antibiotic fails to function effectively by killing the bacteria.

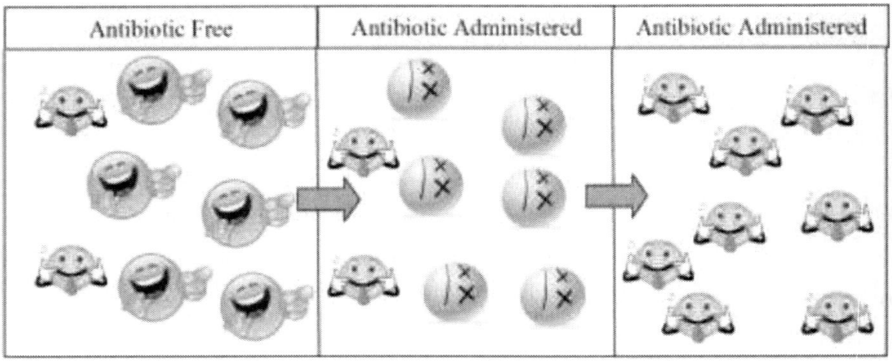

Figure 2.2. Antibiotic Resistance.

There are two known mechanisms where bacteria can develop resistance against antibiotics. The first mechanism is when the bacteria start producing its own antibiotic to protect itself against other bacteria. Some bacteria have been known to have antibiotic producing capability for exactly this purpose. In such cases, the bacteria use its own antibiotic and kill other bacteria making it lowly antibiotic resistance – since it's only resistant to certain types of antibiotics similar to the one it produces on its own. The other mechanism is antibiotic resistance resulting from abuse and overuse. This happens when patients are frequently administered with antibiotics. Such frequent exposure increases the odds for certain bacteria to evolve a high resistance.

Bacteria may develop resistance genetically. This can happen via genetic mutation or by obtaining resistance genetically from other bacteria. Different genetic mutations results in different types of resistance. For instance, some mutations enable the bacteria to eliminate the cells that the antibiotic targets. In other cases, the bacteria produce chemicals similar to enzymes to neutralize the antibiotics. There are instances when the bacterium reacts in such a way that the antibiotic's access point into the cell gets blocked or the antibiotic gets transported far away so that it never reaches its target. Besides genetic mutation, genetic transfer is the other way bacteria achieve resistance. One common gene transfer path is from one generation to the next. The bacteria offspring would then accumulate more resistance traits (via mutation) which are transferred to the next generation. As this process repeats the successive generation of bacteria develops multiple resistance traits and becomes highly antibiotic-resistant. This is known as 'vertical gene transfer' where genes are transferred from parent to offspring. Another gene transfer method is 'horizontal gene transfer' or non-reproductive gene transfer. One way for horizontal gene transfer is via viruses. The viruses act as a genetic transporter between bacteria where the resistant genes from one bacterium is packaged into the virus which it then injects into the new bacterium it attacks. These sorts of bacteria attacking viruses are known as 'bacteriophages'.

TEIXOBACTIN AND THE ICHIP

Since the sixties scientists have developed only two new classes of antibiotics; an addition to twenty two classes developed prior to that. Besides the challenges faced by making antibiotics, the fight against antibiotic resistance has proven to be a losing battle until recently. A team of scientists have discovered a new antibiotic that was found to be non-resistant – it does not develop resistance in bacteria like other antibiotics (Ling et al., 2015; Arias and Murray, 2015; Wright, 2015). This new antibiotic called teixobactin prevents certain types of bacteria from building key parts of their outer cellular coats. Teixobactin could be employed to destroy bacteria that cause tuberculosis and anthrax. The interesting part of teixobactin is that when tested on mice, it was effective in treating antibiotic resistant diseases. Therefore, scientists went one step further by attempting to develop strains of bacteria that may resist teixobactin. The results were negative; showing that teixobactin is significantly resilient to bacterial resistance. This is of course not to say that teixobactin would be permanently resilient to bacterial attack – only that it would take decades before teixobactin – resistant strains of bacteria could evolve. Soil presents the most microbe-rich environment highly suitable for creating antibiotics. Although they thrive in soil, most microbes are difficult to culture well under lab conditions. Thus some researchers invented the iChip. The iChip is a board with some holes on its surface.

The holes in the iChip are filled with soil which is then soaked and shaken in water to release any present microbes. After diluting the sample extensively and mixing it with liquid agar, the agar is poured into the iChip. The sample is diluted to confirm that in each hole holds only a single bacterium. Each hole is plugged by a disc of solid agar. The discs are covered in permeable membrane and inserted into a beaker containing the soil. This way the bacteria have access to all the nutrients in the soil to aid its growth (see Figure 2.3). The idea of the iChip not only offers a method to create the teixobactin antibiotic, it also offers a framework to create a wide range of antibiotics. Twenty five new antibiotic compounds have been discovered using the iChip method. Although teixobactin is noxious to bacteria, it is harmless to mammalian cells. As of now teixobactin has only been tested on mice. Although human trials may take several years and have not yet been performed, the iChip method and the recent discovery of the teixobactin pave a promising pathway towards developing various new types of antibiotics.

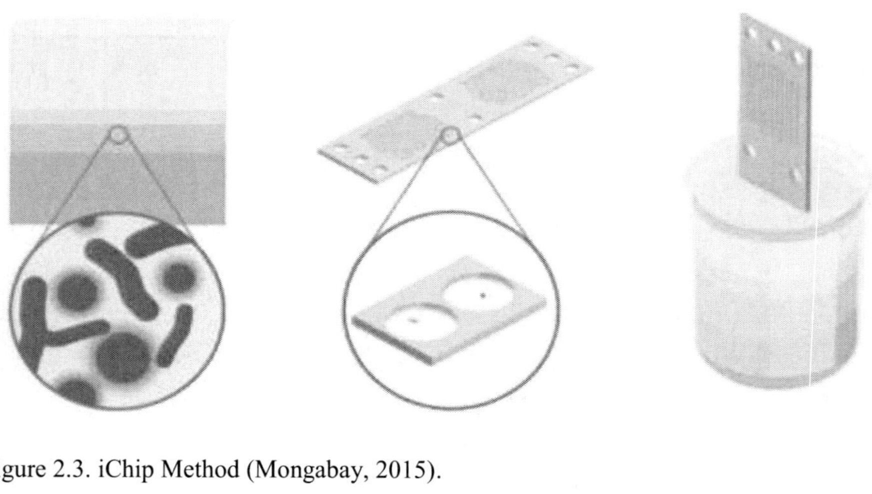

Figure 2.3. iChip Method (Mongabay, 2015).

EVOLUTIONARY ARMS RACE

In addition to antibiotic resistance, epidemics have recently become a global concern. Systemic faults of the market in addition to a growing population is leading to uncontrollable spread of disease – providing a fertile ground for viruses and bacteria to mutate and become resilient. It has to be understood that until now we have cohabited this plan*et al*ong with microbes. The capability for us to evolve alongside them is critical to our continuous existence on the planet. This is to say that we are in an arms race with them in terms of evolutionary rate and capability. The moment our rate of evolution (translating to immunity) drops, viruses and bacteria would quickly have an upper hand (evolving into some kind of superbug). Without cure such superbugs could cause harm on a scale so large it may lead to species extinction.

Among the most effective strategies which could be used as alternatives to antibiotic treatment is passive immunization – where the immunity against microbes is induced artificially by transferring foreign antibodies into the body. One other form of passive immunity (non-artificial) is the antibody transfer from mother to fetus via the placenta during pregnancy. Immunity could also be acquired naturally against certain diseases such as chicken pox and measles – where patients do not get afflicted twice because the immune system works to generate and activate antibodies by itself. This sort of reaction is called active immunity. Although passive immunization is immediately effective, its sustaining power against diseases is short-lived. Note that passive immunity is not the same as vaccination. Vaccination involves the introduction of a weakened form of bacteria/virus into the body to induce a reaction from the immune system. This would then provide the person with immunity from specific viruses/bacteria. Phage therapy is another potential strategy in our battle against antibiotic resistance (Potera, 2013). Phage therapy uses bacteriophages (bacteria killing viruses) to target and eliminate disease-causing bacteria. Phages are extremely specific and since plants and mammals lack receptors required for phage infection, phages do not harm them. Unfortunately, similar to antibiotics, bacteria could acquire immunity to phages while the phages evolve to bypass the bacterial immunity in return – waging an evolutionary battle. Hence some form of resistance to phages may arise and some point in time similar to antibiotic resistance. Nevertheless the option of phage therapy is still being explored as a viable potential to antibiotics.

Antibiotics work by targeting and disrupting cellular processes in bacteria. Current antibiotics target the same processes as their previous versions, e.g., DNA/RNA translation machinery, cellular wall synthesis and DNA/RNA metabolism. Novel antibiotics that target different cellular processes (such as cell division and fatty acid synthesis) are being researched (Su and Honek, 2007; Lock and Harry, 2008). To avoid antibiotic resistance, drug combination therapy has been seen to be more effective compared to singular antibiotic treatment. This is done by mixing the type of therapies e.g., antibiotic and phage therapy. Strategies involving combination therapy has shown to be promising in preventing antibiotic resistance (Lu and Collins, 2009).

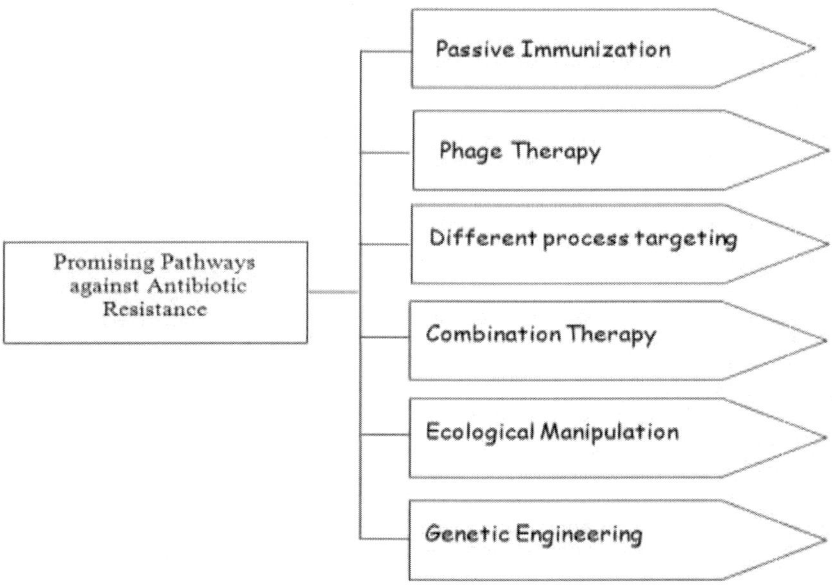

Figure 2.4. Potential Approaches against Antibiotic Resistance.

One potential area for exploration is ecological interactions. Resistant genes develop and commute from one organism to the next. For instance, resistant bacteria obtain their resistant genes from phages which co-evolve and transfer their genetic material with other organisms (e.g., humans or bats) and other viruses. This ecological information transfer needs to be ascertained and understood. In addition to evolutionary and causal ideas, the larger ecological perspective could shed some light on knowledge which may be helpful towards dealing with antibiotic resistance. Tracking down the path at which the resistive genes move would greatly improve our standing. The development of molecular biology and biotechnology could also be used as tools to perform genetic modification as well as modify microbe metabolisms towards reducing or eliminating resistance in disease-causing microbes. Figure 2.4 graphically presents the potential avenues for dealing with antibiotic resistance.

Unlike other organisms, humans have not only developed science but have also evolved under the influence of science. The human physiology has over the years become very dependent on medical technology to survive attacks from the natural environment. Without the advent of some of the developments in medical science such as antibiotics, vaccines and other sorts of drugs, humanity may not be capable of thriving on this planet. To maintain steady scientific and medical development, a dynamic and adaptive research infrastructure needs to be considered. Take the discovery of the iChip. On one hand it could be observed that the iChip is a brilliant discovery in terms of its simplicity and its usefulness. Nevertheless remember that since 1962 scientists have developed only two new classes of antibiotics. After 54 years, only now we've discovered the iChip that could potentially lead to discovering more types of antibiotics.

Something like the iChip does not need other technologies to be discovered first before it could be conceived. For instance computers could not be built without first discovering transistors – this is not the case with the iChip. This begs the question, why was the iChip not discovered in the past 54 years? We should question if our research institutions such as our universities, government and corporate research bodies are efficient, dynamic and adaptive enough to face modern scientific and medical challenges. We should start considering; if the support systems such as the financial, academic and regulatory frameworks we set up to do science will remain relevant and functional in the near future. After all the essence of the scientific method is 'doubt'. So why not cast doubt and look for improvements in the ways at which we do science in the first place.

EPIDEMICS VERSUS MEDICAL EFFORTS

The scientists that developed antibiotics warned about the potential encounter with antibiotic resistance. The emergence of antibiotic resistance presents a huge obstacle to the development of medicines for various diseases. Besides, the slow progress of the development of antibiotics is another major setback faced in medicine. Among the factors that influence the propagation of viruses are populations of their carriers such as birds, swine and rodents. For instance, swine density in farms is a factor which may influence the spread of the swine flu virus (H1N1). Besides, high human population plays an integral role for disease spread (especially virulent ones). Increase in human population causes concerns involving the diminishing of resources (food, fuel, space and water). In the past, pandemics have spread to various densely populated areas such as cities, killing countless in the process. For instance in 1918, the flu pandemic at the end of the First World War killed (tens of millions of people) more people than the war itself (Birch et al., 2013). The 1918 flu virus is a similar strain to the H5N1 bird flu virus that struck the world again in 1997 beginning with an outbreak in Hong Kong. Although the 1918 bird flu spread from person to person, the H5N1 bird flu required birds as an intermediary and could not be transmitted directly among people. The 1997 flu outbreak is a reminder to count our lucky stars because if medical science is not developed enough when the next pandemic (similar to that of the 1918 flu) hits us, then the human population would be devastated on a very massive scale.

As the human population grows, species of viruses have more opportunity to evolve as there is more availability of hosts (Walters, 2004). Why and how does this happen? First let's take a look at how viruses survive. Unlike other organisms, viruses require host cells to survive and reproduce. Viruses do this by attaching themselves to the cell membrane and then injecting its genetic materials (DNA or RNA depending on the virus type) into the host cell. The genetic material contains instructions for creating more viruses and other commands which the host cell will carry out. The new generation of viruses then continues repeating the cycle by seeking out other host cells. The hosts are not passive when something like this happens – the host's immune system quickly picks up and identifies the viral code. Once this is learned by the host's immune system it then develops techniques to prevent reinjection by the virus. That is why certain viral infections like chicken pox and mumps rarely recur in the same person twice. This way the host cell is not very accommodative the next time the virus visits. To survive the virus must mutate and evolve alongside its host so that the host receives the virus' genetic material allowing it to attach, replicate and do its job.

Now think about a small population being infected with a type of virus. With time there would be only some amount of mutations occurring in the virus allowing it to inhabit their hosts. However, with a large population of hosts, the possible combinations at which the virus may mutate would be very high. Dealing with a larger population pool means that the immune system's response of each individual would be diverse as well. Thus the virus would undergo far more mutations at an extremely high rate compared to infecting a smaller population – this makes the virus tougher and more resilient. An excellent example of such a virus is the mosquito-based viral disease, dengue which has rapidly evolved new resilient strains. So much so that viruses such as Zika has made an aggressive come back. The Zika virus is related to dengue, yellow fever, west Nile and Japanese Encephalitis (J.E.) viruses. Extensive medical research and slick antibiotic-generating gadgets such as the iChip will only tackle the symptoms of the problem. The core of disease spread lies in more fundamental issues like population control, poor sanitation, poor access to medication and unavailability of clean drinking water. Take the Zika virus for example. Besides isolated cases in Africa (such as some parts of Senegal and Gabon) and Asia before the nineties, Zika virus was pretty much low-profile and was not a concern. Then a sudden outbreak occurs in Brazil in 2015 which is then followed by a widespread epidemic affecting South America, Mexico and Central America. During the time of writing this book, the Zika virus has already spread into many territories of the United States with 258 travel–associated cases reported.

ECONOMICS OF DISEASE SPREAD

Disease spread quickly transforms from a local epidemic into a threatening pandemic. Policy makers and others always think of disease spread as a gradual process. This is certainly not true. In economics this is called exponential-growth bias – where people think that savings or returns expand linearly with time. This is not the case since in reality interest's compound gradually causing exponential-growth in savings. The same thing happens with disease spread – it does not spread linearly but rather exponentially compounding over a larger population as it grows.

In addition to a growing population, the world is more connected than it used to be – the global community no longer lives in isolation. With readily available transportation, people could easily fly across continents carrying various disease vectors. Besides pumping funds into medical research and boosting healthcare corporations (e.g., pharmaceutical and medical tech–related companies) for economic reasons, the time has come for us to realize that what affects the people in other countries would quickly arrive on our doorsteps in a matter of hours. Although being connected may bring great economic benefit, the responsibility that comes with it needs to be embraced. Scientific efforts should be channeled towards improving sanitary conditions, medical infrastructure and water quality in developing nations. These initiatives could be thought of as part of a long-term epidemic (or pandemic) preventive measure.

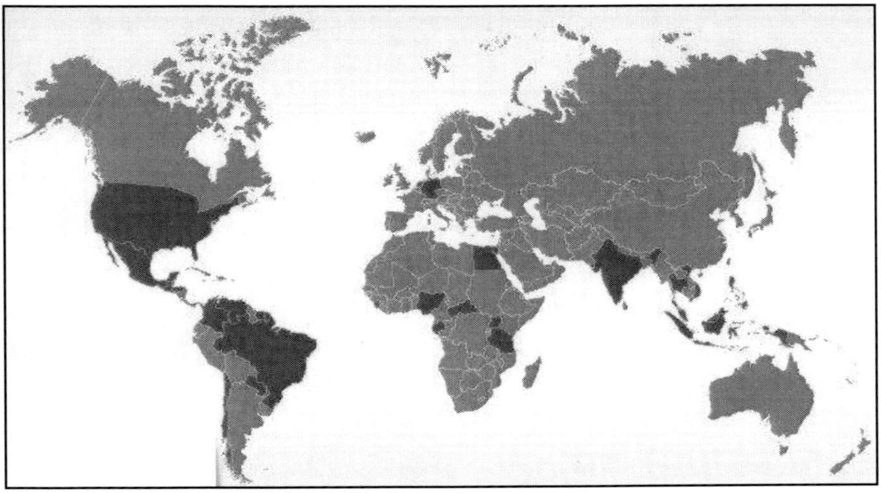

Figure 2.5. Countries with Confirmed Zika cases in Feb 2016 (Daily Star, 2016).

Consider this: if conditions were better in South America, would the Zika virus spread as quickly and as aggressively as it has currently? Figure 2.5 charts out global map of Zika cases confirmed in 2016: Uganda, Nigeria, Tanzania, Egypt, Central African, Egypt, Republic of Sierra Leone, Gabon, India, Malaysia, Philippines, Thailand, Vietnam, Indonesia, Micronesia, Polynesia, Easter Island, The Cook Islands, New Caledonia, Brazil, Chile, Colombia, El Salvador, Guatemala, Mexico, Paraguay, Venezuela, Suriname, Cape Verde, Fiji, French Guiana, Honduras, Puerto Rico, Panama, Vanuatu, USA, Netherlands, Germany (Daily Star, 2016).

At the time this book is being written, the Zika virus has been globally spreading from the Americas to other places in the world. Without a vaccine or antiviral, the virus is freely infecting and evolving among general populations. Our current system of medicine deployment involves large pharmaceutical companies. As a matter of fact, there are only a handful of companies that control global vaccine development. For big pharmaceutical corporations, a reasonable profit developing vaccines could only be made if you have a large enough market – a huge patient pool. Until today there is still not a large enough market to justify making an Ebola vaccine. As said by Professor Adrian Hill from Oxford University who is an expert on vaccine production and evaluation, ‘The problem with that is, even if you've got a way of making a vaccine, unless there's a big market, it's not worth the while of a mega-company' (Krishnan, 2014; Andalo, 2015; Halder et al., 2011; Spalding and Torsoli, 2016). Another critical factor that hinders the economics of developing a new vaccine is that most people affected by the disease are located among some of the poorest countries in the world. Thus they cannot afford to pay for a new vaccine. All this is can be seen as a systemic failure of the global market – where it drops the ball when it comes to critical vaccine development and disease control during an outbreak. I am going to reiterate this because this is very important – the world is now more connected than it used to be where an outbreak of any type of disease has the capability to extend its clutch around the world in a matter of days. Adding to that is the growing world population which offers the microbes a huge enough pool of hosts; adequate for proper evolution making it highly resilient and virulent. We have to remember that disease vectors do not consider factors like corporate profit margins and the affected population size. Therefore, to wait for the sluggish reaction of the market may be detrimental. It is possible that by the time the market reacts, the disease would have already progressed into a very virulent form – making it incredibly difficult to produce a vaccine or antiviral. By that time large casualties would be inevitable.

One way to counter such disease spread is by 'nipping it in the bud' during the initial stages of the outbreak. If our vaccination system is effective and speedy, we could have immediately vaccinated the affected regions during the initial stages of the outbreak crippling its spread. This would require a small quantity of vaccine which is readily available and could be easily deployed to places when affected. A strategy like this would save millions of lives besides saving lots of money spent trying to curb a resilient epidemic (or even worst a pandemic). When it comes to vaccination and disease control we have to admit that at some point the financial system and the free market fails. Besides, as mentioned previously outbreaks are becoming an international issue due its spread. The WHO could partially regulate big pharmaceutical companies (which could be done easily since there are so few of them) such that they would have to develop vaccines and contribute to curbing outbreaks – especially those which are spreading on massive scales threatening a pandemic. This could be seen as a minor social service considering the vast profits these companies make. Here is another way; the WHO and other leading medical research centers could impose a fixed funding from big pharmaceutical companies to enable research into disease control and vaccine development. Not offer research grants but a fixed consistent inflow of cash to enable others to act fast during an outbreak if the pharmaceutical companies find it 'non-profitable'.

No matter how one looks at it, the problem remains – the financial market does not respond well when it comes to vaccine development and disease control. More effort needs to be channeled to the effect that it eliminates the market's inadequacies to seep into critical issues such as disease spread which is increasingly becoming a threat to the global community. If the financial system and market does not function when we need it to, especially when it comes to our own survival, we would have to work our way around it. But first we need to acknowledge that there exists a problem.

Psychological Feedback

In November 2002 a disease known as Severe Acute Respiratory Syndrome (SARS) broke out in the Guangdong province of China. SARS then diversified into 29 regions affecting over 8000 people globally. Subsequently, another outbreak occurred in Hong Kong in the beginning of 2003. SARS then spread to East Asian countries and then towards Canada causing over 770 deaths along the way. SARS is caused by a type of virus called the coronavirus – which upon infection causes flu-like symptoms including fever and cough leading to pneumonia. Thankfully upon successful containment efforts the last known case of SARS was in May 2004 in China (WHO, 2004). The WHO declared that the human-to-human transmission has been broken. Along with the SARS epidemic came the psychological backlash – called 'SARS phobia' (Cheng, 2004). A 48-year-old Taiwanese man committed suicide on April 2003 after suspecting some of his family members have contracted SARS. In Hong Kong a public panic was instigated when someone hacked a local news website and posted a hoax – claiming that Hong Kong would be sealed off from the rest of the world due to widespread SARS infection. Although SARS was not as virulent as some of the other diseases, the psychological anxiety and panic it brought was definitely large in scale. Due to a combination of factors such as; suddenness of the occurrence, the quick death toll and the economic depression at the moment of the outbreak, SARS left the society in those regions in a state of panic and anxiety. It was even said that the SARS psychological effects have been far more severe than the disease (WHO, 2003). During such times, decision making and policy making should be done very carefully. As pointed out by the great sociologist Daniel Kahnemann; the perception of risk and severity of the situation may be distorted (Gilovich et al., 2002). The suddenness and severity of the outbreak would create a strong emotional response from the affected society echoed by policy makers. At the time of the outbreak or immediately after it, sentiments are high and the assessment of risk may be highly biased. This kind of bias is called the affect heuristic. The affect heuristic is a 'mental shortcut' people use (unconsciously) for solving difficult problems quickly. The affect heuristic uses current emotional states for decision making causing solutions and assessments to be highly biased and skewed. It is crucial that decision makers (e.g., government, organizations, scientists and healthcare providers) involved in designing disease mitigation measures to not fall into this sort of bias.

The affect bias is highly prevalent in today's society. A typical example (which you have most probably heard of) is the fear of flying and its associated statistics relating to car crashes. Plane crashes are sudden and highly traumatic. Add that to our innate fear of lifting of the ground and you get a solid bias against flying motivating you to avoid it when you can. If you ask a person to estimate the risk of a plane crash, they would most probably overestimate by a significant margin. Although car crashes may be as traumatic, they happen so frequently that people don't register it with the same emotional strength as compared to plane crashes. This is the sort of bias that prompts people to wrongly assess risk and make highly biased decisions – which often prove to be ineffective in the long run.

The Ebola epidemic infected people living in West African countries in 2014. In the U.S. only 10 confirmed Ebola cases were reported – infecting mostly healthcare workers and people who had direct contact with infected patients. Nevertheless a syndrome called 'fearbola' (as coined by the media) spread across the U.S. The extreme reactions of the American parents were the most striking. If any of the school personnel, teachers or students had gone to Africa or could be somehow connected to people who had contact with Ebola patients, the American parents immediately pulled their kids out of the associated school (Lu, 2015). It went so far that even a Maine school teacher who attended a conference in Dallas (where Ebola first broke out in the U.S.) was put on leave from work. These sorts of scares are very much associated with the idea of how we perceive and assess risk. In the words of Paul Slovic, a Professor at University of Oregon on the 'fearbola' situation – 'What happened was quite consistent with what we know about risk perception. The minute the Ebola threat was communicated, it hit all of the hot buttons: It can be fatal, it's invisible and hard to protect against, exposure is involuntary and it's not clear that the authorities are in control of the situation.'

Similar to the shocking plane versus car crash statistics, influenza causes 20 percent of the global population to get sick. Among those, thousands die annually. Due to the frequency of infection and the amount of people who usually recover from the disease, it is not perceived as a threat and its risk is undervalued. That is the reason why most people don't see the importance of vaccinating themselves against influenza. Diseases which are new, fatal and brutal cause a more strong fear reaction from most people and are perceived as a greater risk. Here comes affect bias again clouding our judgment and assessment of the situation. It gets really difficult to circumvent affect bias if someone close to you; such as a family member or a loved one gets affected. No one can blame you for being biased – that sort of thing would cloud any of our judgments. Otherwise the best way to overcome affect bias is to ask reasonable questions about the situation and extract as much data as possible. For instance, identify how the disease spreads and what its spreading rate is like. Then ascertain how many people are affected and in what regions. Try to move your focus away from the suffering and other things that invoke strong emotions. Simply put things in perspective and look at the data as objectively as possible. As has been seen often in science, objectivity really pays off. This way you could review the data with a clear mind and assess the potential threat with an accurate risk assessment; free from the affect bias.

The other bias that heavily influences our assessment of risk is the 'framing effect'. The framing effect occurs when people interpret information based on how it's presented. For instance; what if the news on Ebola is presented as a short segment between other more 'critical' issues such as; terrorism and political upheavals around the globe. Then the severity of the infection risk would definitely be underrated. In fact if it was presented as another version of influenza-like disease, then it would most surely be considered as commonplace and taken with similar caution as other issues. What if you provide a half an hour segment on the outbreak during prime time television. To top that, other news flashes are added and labeled as 'breaking news' (like they always do) between your favorite television programs. That would definitely raise panic and instigate a fear response that would permeate throughout society. This would in effect prompt government representatives and other policy makers to develop immediate mitigation strategies and so on while efforts and resources could have been channeled to handle other more urgent issues. Acting on biased assessments is a waste of resources. Besides the media, the framing effect could get to you from unreliable internet sources. For instance; even while writing this book I closely checked the sources from which I extract my information to confirm its credibility and shield me from drawing biased opinions on the issues discussed. Since the advent of independent bloggers, information has become abundant and often emotionally charged. This is exactly the sort of thing that can cause people to cast judgments and opinions under the influence of the framing effect. On the internet, information or statements often go 'viral' – it should be understood that this does not make it true or accurate. In fact, I advise readers of this book to look up some the issues I have highlighted and checkout some of the references I have included in this book. This way you could get an unbiased and clearer picture of the issues presented here. In my view the threats posed by diseases and pandemics are real. Our efforts to prevent future threats are highly necessary though they should be carried out clear-mindedly with accurate data along with good planning – which is free from biases.

LOST IN PUBLICATION

Medical research progresses similarly as to other branches of science – where advancements and recent developments are published in 'peer reviewed' journals (or other avenues that disseminate scientific information). The peer review system greatly influences the pace at which medical science progresses. This system also impacts the reliability of medical knowledge being distributed among medical researchers. In this section we will explore how peer review gets factored into medical developments and see how it affects modern day science. Utilized in research publication, peer review is a checking mechanism serving as the gatekeeper for quality assurance in science. Here is how it works: first a scientist performs a research setting out to prove or disprove some hypothesis (or scientific question). Upon completion, the scientist(s) starts the writing a report to publish the findings of the research. Customarily such findings would be submitted to a scientific journal (or conference). Before acceptance for publication, the editor would screen the paper to see if it is relevant or within the scope of the journal. Besides the editor also performs a quality check at first glance to see of the submission lives up to the editor's expectations in terms of quality. Once this stage is cleared, the editor usually sends the submission to two reviewers who are usually 'peers' in a sense that they are researchers work on a similar field with the author. Therefore, these two reviewers would evaluate the work either agreeing or disagreeing to accept the work for publication. The reviewers may also conditionally accept the submission requiring the authors to perform some corrections and modification to the original work prior to publication, often the case with most scientific publications. If there happens to be a disagreement between the two reviewers, then a third reviewer is called in to finalize the decision. Therefore, only articles that meet high scientific standards are published. The reasoning behind peer review is that only high-quality scientific work is published. Hence scientists are able to base their work on strong research findings. Since scientific work is largely cumulative, this leads to the production of reliable scientific work.

There are predominantly three types of peer review; double-blind, single-blind and open. In a double-blind review, the identity and background of both authors and reviewers are not revealed to each other. On single-blind reviews, the identity of the author is known to the reviewer but not the way around. Both party identities are known to each other in open reviews. Clearly the most non-biased system would be double-blind reviews since both parties have no knowledge of each other's identity. Peer review is hoped to be a non-partial, reliable, objective and consistent framework for checking what passes as scientific work. Today peer review has been used extensively in various aspects of science and medicine (Lee et al., 2013):

- Observation of peer's clinical practice
- Research grant approvals
- Assessment of colleagues teaching abilities in university lectures
- Book proposal evaluations
- Consideration for academic promotions in universities and the awarding of tenured positions.

Despite its widespread usage, peer review continues to attract many criticisms – as we shall see here some of these critics are not without ground. The primary problem with peer review is that there is no agreed idea on what constitutes a 'good paper', a 'good grant proposal' or a 'good conference presentation'. This is where the review starts to become very subjective. Remember we are discussing the starting point of science – where ideas are thrown around and experimented with. In these preliminary stages; concepts and frameworks are being tested and analyzed, research proposals are being written and ideas are presented at conferences. None of these concepts are proven or functional yet. This is why it's very difficult to distinguish good ideas from bad ones. Researchers have found it very hard to determine which papers should be published and which should be rejected. Robbie Fox, a distinguished editor of the eminent British medical journal, Lancet stated; that no one would know if he swapped the labels of papers marked 'publish' or 'reject'(Smith, 2006). This is because peer review could also be highly inconsistent making the results no different than an outcome of a game of chance.

Experts in many scientific disciplines are inclined towards a school of thought – which they think is the way forward in their respective fields. Each of these experts may have developed different schools of thoughts which could conflict with one another. Here peer review becomes an obstacle since if the other reviewer (or expert) subscribes to a different school of thought, then many other ideas would never see the light of day. If this reviewer is eminent and becomes an editor for major scientific journals, this could significantly hamper the development of the field since many novel ideas would be rejected at first glance. The situation could worsen if such experts are on grant approval committees. Not only would it be difficult to publish novel ideas, it would also be very difficult to secure grants for novel scientific research.

One interesting example is the cause for Alzheimer's disease (albeit remains unknown to this day). Medical researchers have fallen into two groups subscribing to two schools of thought. One group thinks that Alzheimer's is most probably caused by aggregation of substances called tau proteins while the other pursues ideas stating that Alzheimer's is caused by another protein called amyloid beta. The two groups are often referred to humorously as the Taoist and Baptists of Alzheimer's research (Mudher and Lovestone, 2002). Another instance where scientists form camps is in physics. Yes, in physics where in seeking for a unified theory of physical world they too fall into groups such as loop quantum gravity and superstring theory. Therefore, any physicist that may have a completely novel approach to unification that does not subscribe to any of the above so called 'prominent' ideas can forget about grants and publications – in essence forget about a career in academic research. There are many such examples in most areas in science. This causes research funds and publications to be centered around certain elite groups killing off the competition – which is any other researcher which does not follow certain prominent views. As more scientific fields emerge and more researchers get involved in the pursuit of science, 'grouping' tendencies become more prevalent. Like other human institutions; universities, academic journals and research groups are no exception. Such groupings could be very dynamic and productive; especially when scientist could explore various hypothesis and research avenues simultaneously cutting down the research timeframe by acquiring knowledge swiftly. On the flip side, things like schools of thoughts, and conflicts from competition for access to limited resources (such as funding and academic publication) start to weaken the scientific enterprise.

Besides favoring like-minded colleagues and certain groups, other types of biases seep into peer-review. Despite many women practicing science in the last century, science still remains a male-centered domain (Kaatz et al., 2014). Research has shown that science is often associated with the male gender – hence leaning more in favor of them. A randomized experiment was designed by researchers to study the effects of gender bias when hiring a laboratory manager. The results showed that the odds leaned towards male applicants – where the employers would rather hire a male applicant; providing him with career training and paying him a higher wage as compared to a female applicant (Moss-Racusin, 2012). In one peer-review test; the reviewers were given an implicit association test to identify their biases. Studies showed that reviewers in the age groups of 40 to 65 had the strongest association of male and science (Nosek et al., 2007). In the seventies another study was performed on a group of students; where the students were randomly asked to read a same article written by the author with either the first name 'John' or 'Joan' (Goldberg, 1968). The students were then asked to rate their respective articles. In spite of being an all-female group, these students gave higher scores to the articles written by John than to the one by Joan. Statistically, female researchers in the U.S. between the years 2003 and 2007 have shown to obtain considerably lower RO1 funds as compared to their male counterparts (RO1 is a National NIH medical grant funded by the U.S. government). Besides having lower publication rates than male scientists, women are also often have little success when it comes to applying for research grants.

Accentuating the Positive

Despite the fact that science has been practiced since the times of Galileo and Copernicus, it still has to jump various hurdles to progress in a swift, efficient and not to mention ethical manner. The tendency of not publishing negative results is another thing that greatly slow down the machinery of science. What are negative results? When we begin scientific exploration, we first construct a hypothesis – which is more like a possible solution to the some question. This hypothesis can either be proven or disproven. Both cases being invaluable in its own respect since it paves the way forward taking us closer to the truth or the true solution. Negative results are when the hypothesis is disproven. For some reason such results are often deemed unsuitable for publication. It seems like researchers would rather read about the success stories and not the reverse. This is strange because the hypotheses could always be reframed to prove it rather than disprove it. Consider this example hypothesis; influenza is caused by a type of bacteria. After scientific testing, this hypothesis would be disproven obtaining negative results. This is a good thing since now we will continue research finding for the true cause of influenza – eventually discovering the rhinovirus. The hypothesis could also be proven if we reframe it as follows; influenza is *not* caused by a type of bacteria. In this case we would prove it and continue our research finding influenza's true cause. The point here is that negative or positive results are equally crucial in science. Negative results tell other researchers which research directions are futile. This in return saves considerable amount of time and resource which would be wasted on redundant research. Besides unlike other fields, science has always been known to present all sides of story in the most objective style.

Scientific progresses steadily as ideas are generated. The ideas then become hypotheses and subsequently they are either proven or disproven. This idea is then modified, changed or discarded as time goes by. This way science steadily gravitates towards the truth. Although the scientific process works this way most of the time, there are staggering exceptions. Take for instance the case of the medical researcher, Andrew Wakefield in late nineties. Wakefield along with 12 other co-authors published an article claiming that Measles, Mumps and Rubella (MMR) vaccination increases the chances of child autism (Rao et al., 2011; Matosin et al., 2014). Appearing in the prestigious journal, *Lancet*, the article started an unprecedented panic reaction. For almost a decade, the rate of child immunizations was in the decline. Some thirteen research works producing negative results were published leading in the retraction of Wakefield's article from *Lancet* in 2010. Although its infamy became widespread since its publication in 1998, the impact of Wakefield's article was still influencing the decision of the masses against vaccination. This made diseases such as MMR on the rise again leading to many deaths worldwide. Today remnants of Wakefield's ideas could still be seen surviving in the current Anti-vax movements (Novella, 2016). This is one instance where although the hypothesis has been tested and proven false, its initial impact still lingers around.

From the time it was published it took the Wakefield article about ten years to be retracted. Once published, it gets very difficult to correct or retract scientific papers. The Australian cell biologist David Vaux was reviewing a journal article on the discovery of a method to prevent the rejection of transplanted tissue (Bellgrau et al., 1995). Organ transplant patient often reject transplanted organs and tissues for various reasons. The article reviewed by Vaux was describing a method to avoid this rejection – which was great since this would make organ transplants to sick much more feasible. Influenced by the article, Vaux continued the research in his own laboratory. However, he soon found that he couldn't reproduce the results as the initial article – all he got was negative results. Vaux then made failed attempts to publish his negative results in the journal, *Nature* and *Nature Medicine*. Despite the difficulty faced by Faux in correcting the errors appearing in the published paper, he finally published his negative results after two years in *PNAS* (Allison et al., 1997).

Research quality as well as efficient developments in medicine could only progress if the whole scientific process is self-correcting. Self-correction is only possible if negative results could be easily published and accepted. Scientists should be able to correct published articles. Finally if it comes to be known that the article is erroneous, then it should be retracted – so that others do not continue building their research on mistaken results. Scientific achievement is not the only motivation of doing science. We should always remind ourselves that science is the pursuit of truth – not the preservation of the status quo or reaching desired results. Whether desired or not; we should analyze the results from our experiments and work towards building a consistent framework by filling information gaps. This could only be achieved if the scientific enterprise as whole functions accordingly.

One approach to reduce bias in peer–review is to make science more available and transparent for correction. This could be done by conducting a very preliminary or light review before publication – to ensure that the author performs the scientific method satisfactorily and abides to other formalities such as scientific notation. Thus the reviewer minimally corrects the author and the article is quickly published. Upon publication, the article is opened to other researchers. In time the scientific masses (and not some 'eminent expert' in the field) could determine if the article is to be modified, corrected or discarded. The article would then be scrutinized by the scientific masses; 'mass review'. Besides ensuring overall scientific quality via consistent correction, the whole process becomes significantly transparent circumventing scientific elitism, groupings and biases.

DEMOCRATIZING SCIENCE

The thing about publication biases is that it gradually leads to censorship – where only dominant scientific ideas and works are publicized while others are left to wither away to obscurity due to censorship. A critical step against such biases was taken by Paul Ginsparg in 1991 where he initiated what is today known as arXiv (pronounced 'archive'). ArXiv is an online repository for preprints in physical sciences, mathematics and computing. ArXiv is hosted and maintained by Cornell University (Ginsberg, 2008). A preprint (also known as e-print) is a draft of a scientific paper which has not undergone rigorous peer-review in any scientific journal. Since arXiv is open-sourced, preprints uploaded there could be easily downloaded for free by researchers around the globe. Therefore, arXiv significantly reduces biases by making publication easy without requiring rigorous peer-review. Besides, it increases transparency since the research community could easily download articles from arXiv for free.

In most cases, preprints published in arXiv are also published in other reputed journals. To maintain certain standards in terms of the quality of submissions, a group of volunteer moderators (being experts in their respective fields) were put in place. These volunteer moderators screen through the submissions deciding if the article meets the minimal quality requirements to be uploaded on arXiv. Those that do not meet the standards set by these moderators are rejected. All first time authors are required to be endorsed by past, long time arXiv contributors. Those that do not get endorsements would not be able submit. Unfortunately all this sounds like regular a peer-review journal all over again.

Consider this rather dramatic example; if Einstein being a patent clerk was to discover the theory of relativity this time around and proceed to submit it to a preprint repository (or any other journal) – he would probably face rejection. For two reasons; first, besides being theoretical, his idea is radically new and unheard of. This would merit rejection by the volunteer moderators/editors. It would not be surprising if comments like 'not experimentally validated', 'lack of basis', 'unproven' and 'groundless claims' appearing in the review report. Being a patent clerk without academic affiliations would make finding potential endorsers next to impossible. In scientific publishing, prominence and academic status often gets confused with credibility – which leads to biased screening or review. Unfortunately many scientists with really great ideas like Jenner and Pasteur have dealt with difficulties in publishing their work in the past due to reasons mentioned above. The question we should ask ourselves is; do online preprint publishers like arXiv help our problems with publication bias and censorship? As with conventional peer-review journals, preprint systems would work better if the articles are made easier to be retracted, corrected or modified post-publication. With such a mechanism; after a simple preliminary screening by volunteer moderators, the article could then be let out to be scrutinized by other researchers. Based on various feedbacks and other indicators (such as citations and reproducibility), the article could then be corrected or removed from circulation. This way a more democratic and transparent process is put in place of regular peer-review to discourage bias-filled scientific elitism. In arXiv, many moderators in fields like mathematics and physics remain unknown to the public (arXiv, 2016).

In 2001, Robert Gentry, Seventh-day Adventist filed a lawsuit in Tennessee against arXiv for removing his papers and revoking his posting rights (Brumfiel, 2008). He claimed that they rejected his articles due to their religious content. Holding a Master's degree in physics, Gentry has written papers in nuclear geophysics published in well acclaimed journals. His research then inspired him to develop an alternative to the Big Bang theory. Facing rejection from many academic journals he proceeded to publish it in arXiv. After receiving the cold shoulder from arXiv administrators he sued stating, "I'm a creationist and a believer in the Bible, but I want to know the truth. I want these papers to be tested by the scientific community." Some scientists have taken up roles as guardians of knowledge – deciding which passes as science and which does not. This is the first few steps of how censorship begins. There is no reason for any scientific article or idea to be halted from publication. Similarly such ideas should be open to scientific testing. There shouldn't be any authority deciding which ideas should be tested and which should not. Many scientists supporting such screenings hide behind the banner of 'quality control' – where of course they have the final word on what 'quality' is.

The emergence of scientific elitism, peer-review biases, resistance towards negative results and academic censorship are among the factors slowing down the progress of science and medicine. Scientific knowledge is obtained gradually over many research works. The complete framework of ideas is usually pieced together over time. Instead of rigorous screening, evaluation and rejection; why not let research work be corrected gradually over time by the scientific community at large. As mentioned before, a preliminary screening would be sufficient. The publication could then be left to be scrutinized by other researchers eventually deciding its fate; to be modified, left as it is or removed from circulation. Any scientific concept is not completely understood in its first encounter. It's rather understood gradually over time as we get increasingly exposed to other information/ideas – then we eventually see the 'bigger picture' of how the ideas relate to each other. That is how we form a stronger understanding of scientific ideas. This process of continuous correction should be adopted into scientific publication and research to smoothen and fasten scientific and medical efforts. Although seemingly unrelated, our methods of practicing science do impact the speed at which we create vaccines and other treatments to curb disease spread.

Similar to arXiv, a preprint server for biology was developed in 2013 by Cold Spring Harbor Laboratory (CSHL) (Kaiser, 2013). As for medicine, one recent avenue for open peer-review preprints is PeerJPrePrints. Preprint servers are currently giving us an additional edge when battling epidemics such as the one caused by the Zika virus. Speeding-up the dissemination of current research without going through the lengthy peer-review process enables medical researchers to expedite research towards finding a vaccine against the Zika virus (Greenwood, 2016). These new preprint servers are moving us towards a more democratic, efficient and bias-free way of doing science and medicine.

In Brief

Antibiotics have been prevalent in ancient cultures and civilizations e.g., Egyptian, Nubian and Chinese. Thanks to scientists like Alexander Fleming, Paul Ehrlich and Tu Youyou, antibiotics have been systematically applied in modern medicine for helping millions around the world fight infectious disease. Unfortunately antibiotic resistance looms with increased usage of antibiotics. Medical research breakthroughs such as the iChip aim to generate antibiotics which oppose resistance. An ecological perspective on resistant genes may be a promising avenue for furthering our understanding on antibiotic resistance – giving us more ideas on how to beat it. The financial system, international organizations and the growing population make it difficult to control the global spread of epidemics. Measuring the severity of disease spread as well as assessment of its risks could be done more effectively by; 1) ensuring bias-free analyses and 2) maintaining objectivity during the observation of the situation. Issues such as publication biases, academic censorship and setbacks of the peer-review system significantly impacts progress in science – specifically medical research. These issues negatively affect the speed and efficiency at which we could react to epidemics and other disease. The new preprint servers are effectively tackling some of these issues.

References

Allison, J., Georgiou, H.M., Strasser, A. and Vaux, D.L., 1997. Transgenic expression of CD95 ligand on islet β cells induces a granulocytic infiltration but does not confer immune privilege upon islet allografts. *Proceedings of the National Academy of Sciences*, 94(8), pp.3943-3947.

Aminov, R.I., 2010. A brief history of the antibiotic era: lessons learned and challenges for the future. *Frontiers in microbiology,* 1, p.134.

Andalo, D., (2015). Cost of vaccines becoming unaffordable in developing countries, charity warns, *Pharmaceutical Journal,* Royal Pharmaceutical Society Publication.

Arias, C.A. and Murray, B.E., 2015. A new antibiotic and the evolution of resistance. *New England Journal of Medicine,* 372(12), pp.1168-1170.

Bassett, E.J., Keith, M.S., Armelagos, G.J., Martin, D.L. and Villanueva, A.R. (1980). Tetracycline-labeled human bone from ancient Sudanese Nubia (A.D. 350). *Science* 209, 1532–1534.

Bellgrau, D., Gold, D., Selawry, H., Moore, J., Franzusoff, A. and Duke, R.C., 1995. A role for CD95 ligand in preventing graft rejection. *Nature,* 377(6550), pp.630-632.

Bennett, P.M., 2008. Plasmid encoded antibiotic resistance: acquisition and transfer of antibiotic resistance genes in bacteria. *British journal of pharmacology,* 153(1), pp. 347-357.

Birch, H., Looi, M.K., & Stuart, C. (2013). *The big questions in science: The quest to solve the great unknowns.*

Brumfiel, G., (2008). Ousted creationist sues over website, *Nature* 420, 597 (2002).

Chain, E., Florey, H.W., Gardner, A.D., Heatley, N.G., Jennings, M.A., Orr-Ewing, J. and Sanders, A.G. (2005). The classic: penicillin as a chemotherapeutic agent. 1940. *Clinical Orthopaedics and Related Research,* 439, pp 23–26.

Cheng, C., The psychology behind the masks: Psychological responses to the severe acute respiratory syndrome outbreak in different regions, *Asian Journal of Social Psychology,* Vol 7, pp 3 -7.

Cook, M., Molto, E. and Anderson, C. (1989). Fluorochrome labelling in Roman period skeletons from Dakhleh Oasis, Egypt, *American Journal of Physical Anthropology,* 80, 137–143.

Cui, L. and Su, X.Z. (2009). Discovery, mechanisms of action and combination therapy of artemisinin. *Expert Review of Antibiotic Infection Therapy* 7, pp 999–1013.

Dharmananda, S., (2002). *Ching-Hao and the Artemisias Used in Chinese Medicine.*

Enne, V.I., Bennett, P.M., Livermore, D.M. and Hall, L.M. (2004). Enhancement of host fitness by the sul2-coding plasmid p9123 in the absence of selective pressure. *Journal of Chemotherapy,* 53, pp 958–963.

Greenwood, V., (2016). New Insights Into How Zika Harms the Brain, *Neuroscience,* Quanta Magazine.

Gilovich, T., Griffin, D. and Kahneman, D., 2002. *Heuristics and biases: The psychology of intuitive judgment.* Cambridge University Press.

Ginsparg, Paul. Winners and Losers in the Global Research Village. T*he Serials Librarian.* 3 (30): 83-95. Retrieved July 6, 2008.

Halder, N., Kelso, J.K. and Milne, G.J., 2011. Cost-effective strategies for mitigating a future influenza pandemic with H1N1 2009 characteristics. *PLoS One,* 6(7), p.e22087.

Hughes, J.P., Rees, S., Kalindjian, S.B. and Philpott, K.L., (2011). Principles of early drug discovery. *British journal of pharmacology,* 162(6), pp.1239-1249.

Kaatz, A., Gutierrez, B. and Carnes, M., 2014. Threats to objectivity in peer review: the case of gender. *Trends in pharmacological sciences*, 35(8), p.371.

Kaiser, J., (2013). New Preprint Server Aims to Be Biologists' Answer to Physicists' arXiv, *Science.*

Krishnan, J., (2014). *Ebola outbreak: Why has 'Big Pharma' failed deadly virus' victims?,* Independent.

Lee, C.J., Sugimoto, C.R., Zhang, G. and Cronin, B., 2013. Bias in peer review. *Journal of the American Society for Information Science and Technology,* 64(1), pp.2-17.

Ling, L.L., Schneider, T., Peoples, A.J., Spoering, A.L., Engels, I., Conlon, B.P., Mueller, A., Schäberle, T.F., Hughes, D.E., Epstein, S. and Jones, M., 2015. A new antibiotic kills pathogens without detectable resistance. *Nature,* 517(7535), pp 455-459.

Lock, R.L. and Harry, E.J. (2008). Cell division inhibitors: new insights for future antibiotics. *Nature Reviews Drug Discovery,* 7, 324–338.

Lu, S., (2015). An Epidemic of Fear, *American Psychological Association,* Vol 46, No 3, pp 46 – 47.

Lu, T.K. and Collins, J.J. (2009). Engineered bacteriophage targeting gene networks as adjuvants for antibiotic therapy. *Proceedings of the National Academy of Sciences*, U.S.A. 106, 4629–4633.

Mahoney, J.F., Arnold, R.C. and Harris, A.D., 1943. Penicillin Treatment of Early Syphilis—A Preliminary Report*. *American Journal of Public Health and the Nations Health,* 33(12), pp.1387-1391.

Matosin, N., Frank, E., Engel, M., Lum, J.S. and Newell, K.A., 2014. Negativity towards negative results: a discussion of the disconnect between scientific worth and scientific culture. *Disease Models and Mechanisms,* 7(2), pp.171-173.

Moss-Racusin, C.A., Dovidio, J.F., Brescoll, V.L., Graham, M.J. and Handelsman, J., 2012. Science faculty's subtle gender biases favor male students. *Proceedings of the National Academy of Sciences*, 109(41), pp.16474-16479.

Mudher, A. and Lovestone, S., 2002. Alzheimer's disease–do tauists and baptists finally shake hands? *Trends in neurosciences,* 25(1), pp.22-26.

Nelson, M.L., Dinardo, A., Hochberg, J. and Armelagos, G.J. (2010). Brief communication: mass spectroscopic characterization of tetracycline in the skeletal remains of an ancient population from Sudanese Nubia 350–550 CE. *American Journal of Physical Anthropology.* 143, pp 151–154.

Novella, S., 2016. The anti-vaccination movement. *Skeptical Inquirer,* 31, pp.26-31.

Nosek, B.A., Smyth, F.L., Hansen, J.J., Devos, T., Lindner, N.M., Ranganath, K.A., Smith, C.T., Olson, K.R., Chugh, D., Greenwald, A.G. and Banaji, M.R., 2007. Pervasiveness and correlates of implicit attitudes and stereotypes. *European Review of Social Psychology,* 18(1), pp.36-88.

Ochman, H., Lawrence, J.G., & Groisman, E.A., 2000. Lateral gene transfer and the nature of bacterial innovation. *Nature,* 405(6784), pp. 299-304.

Potera, C., 2013. Phage renaissance: new hope against antibiotic resistance. *Environmental Health Perspectives,* 121(2), p.a48.

Rao, T.S. and Andrade, C., 2011. The MMR vaccine and autism: Sensation, refutation, retraction, and fraud. *Indian journal of psychiatry,* 53(2), p.95.

Shouzhong, Y., (1998), (translator), *The Divine Farmer's Materia Medica,* Blue Poppy Press, Boulder, CO.

Smith, R., 2006. Peer review: a flawed process at the heart of science and journals. *Journal of the royal society of medicine,* 99(4), pp.178-182.

Spalding, R. and Torsoli, A., 2016. *Zika Fears Spur Pharmaceutical Gold Rush Amid Vaccine Vacuum, B*loomberg.

Su, Z. and Honek, J.F. (2007). Emerging bacterial enzyme targets. *Current Opinion in Investigational Drugs,* 8, 140–149.

Walters, M.J., 2004. *Six modern plagues and how we are causing them.* Island Press.

Wen J.M. and Seifert G., 2000. *Warm Disease Theory, Paradigm Publications,* Brookline, MA.

World Health Organization (2003). *Severe acute respiratory syndrome (SARS): Status of the outbreak and lessons for the immediate future,* Geneva, Switzerland, World Health Organization.

Wright, G., 2015. Antibiotics: An irresistible newcomer. *Nature,* 517(7535), pp.442-444.

Youyou, T., (2011). The discovery of artemisinin (qinghaosu) and gifts from Chinese medicine, *Nature Medicine,* 17 (11), pp 1217 – 1220.

Web Sources

Alliance for the Prudent use of Antibiotics (APUA), 2014. <http://www.tufts.edu/med/apua/about_issue/about_antibioticres.shtml >arXiv (2016). http://www.newworldencyclopedia.org/ entry/ ArXiv#cite_ note-2.

Daily Star – (2016). http://www.dailystar.co.uk/news/latest-news/ 492577/Zika-virus-will-hit-Summer-holidays-Mediterranean-south-america-brazil-mosquitos-transmit

Drug Resistance – Centre of Disease Control (CDC), 2016. <http://www.cdc.gov/drugresistance/>

European Centre for Disease Prevention and Control/European Medicines Agency Joint Working Group (ECDC/EMEA). (2009). *The Bacterial Challenge: Time to React.* <www.ecdc.europa.eu/en/publications/Publications/0909_TER_The_Bacterial_Challenge_Time_to_React.pdf.>

Phenomena National Geographics, 2015. http://phenomena.nationalgeographic.com/2015/01/07/antibiotic-resistance-teixobactin/

Mongabay, 2015. <http://news.mongabay.com/2015/07/antibiotics-revolution-ichip-promises-resistance-free-antibiotics/>

WHO, 2004. http://www.who.int/csr/don/2004_05_18a/en/

Zika Virus – Centre of Disease Control (CDC), (2016), < http://www.cdc.gov/zika/geo/>

Chapter 3

ACCOUNTING FOR GENES

"DNA neither cares nor knows. DNA just is. And we dance to its music."

– Richard Dawkins

ANSWERING THE ULTIMATE QUESTION

In the early 1940s, a series of lectures on biology were delivered by the renowned scientist, Erwin Schrödinger at Trinity College, Dublin. The mention of the name Schrödinger quickly brings to mind the strange quantum physics paradox – involving the survival of cats in boxes. Schrödinger was not a biologist but rather a physicist who built the foundations of quantum physics alongside giants such as; Albert Einstein, Neils Bohr, Max Planck and Werner Heisenberg. In the Trinity College lectures, Schrödinger discussed the physical basis of life – the bolts and nuts which make-up the machinery of life on earth. He tried to explain the workings of biological life using our understanding of physics and chemistry. The following year, the contents of the lectures became one of the most influential science book of our time entitled 'What is Life' (Schrödinger, 1944). In his efforts to link biology with the physical sciences, Schrödinger introduced some revolutionary ideas that will alter and shape our perspective of life.

At the time, the Deoxyribose Nucleic Acid (DNA) was not yet recognized as the substance consisting of genes – it was thought that hereditary materials were carried by certain proteins. Schrödinger argued that hereditary material would have to be an aperiodic crystal. The idea is that solid crystals have a periodic or repetitive structure, and with time could grow by making more similar structures. He reasoned that biological life may similarly function like crystals – where instead of a periodic crystal structure, an aperiodic crystal structure (non-repetitive) sits at the core of all living cells. He proposed non-repetitive crystals because life is a bit more complicated than regular crystals – even a single cell has many components and sub-components which perform many functions. To account for all these functions an aperiodic crystal would be more capable to develop into the multitude of components required to perform various functions. This structure then develops by growing more aperiodic structures. Schrödinger thought that this is the basis of all life forms. When organisms reproduce, this aperiodic structure is carried forward into the next generation causing the crystal to continue replicating growing the new organism. This aperiodic crystal is a sort of 'code script' which contains all of the organism's future growth plan and functioning. Schrödinger thus became the first to propose the idea that organisms function using instructions from a set of codes stored in an aperiodic crystal which is then carried forward into the next generation. Inspired by the book 'What is Life', the molecular biologists Francis Crick and James Watson focused their studies using X-ray crystallography to search of this 'aperiodic' crystal (today known as DNA). The 'code script' concept draws parallels between the functioning of life forms and that of computers – remember this is all taking place in the forties when computing was at its infancies, barely out of the drawing boards.

Around that time another mathematician, John Von Neumann started thinking along the same lines – linking genetic information and computers. Von Neumann considered the gene as a 'tape' encoded with information. This tape has the capacity to program the organism accordingly similar to the operation of a computer (at the time was newly discovered by Alan Turing which he called 'Turing Machines'). The revolutionary idea of taking a computer perspective when thinking of life forms opened up a whole new field of study called cybernetics. Von Neumann was so impressed with the capability of computers he stated that there's nothing computers could not achieve if instructed precisely. In his own words; 'If you tell me precisely what it is a machine cannot do, then I can always make a machine which will do just that' (Atlantic, 2007). Extremely inclined with computers and the way we think, Von Neumann wrote the book – The Computer and the Brain (Von Neumann, 1958). Schrödinger was trying to use the concept of computers to help explain the inner workings of biological life and how genetic information is transferred. Von Neumann on the other hand was using the idea of computers to uncover how the brain functions.

Another key idea presented in the Trinity College lectures was how life forms tend to reverse the notion of entropy encountered in physics. Entropy could be thought of as a useful notion used to measure disorder. Low entropy values means the system is quite ordered (pretty well-organized) while high values of entropy reflects higher degrees of disorder. In most known processes the degree of disorder usually increases as the process progresses. For instance; gases tend to spontaneously expand (and not compress) achieving higher degrees of disorder in its molecules. Another example; left unaided, heat tends to move from high temperatures to low temperatures (not the reverse) progressing towards greater levels of disorder. However, Schrödinger pointed out that this is not the case with life forms. Living organisms have the capability of reversing entropy creating order from disorder – tendencies to grow and reach higher degrees of order, self-organizing its way throughout its life. Organisms do not degrade away like other non-living things but rather they self-sustain by maintaining order in themselves via organizing, growing and repairing themselves – in other words, metabolizing.

Schrödinger went even further by introducing the concepts that employ 'quantum physics to uncover the mysterious nature of genes. During this time he was very much influenced by the thoughts of one German-

American physicist – Max Delbrück. Although a physicist, like Schrödinger Delbrück was interested in figuring out how biological systems function – what makes life tick. His research centered on the study of mechanisms involving bacteria and the viruses that attack them – winning him the 1969 Nobel Prize in Physiology or Medicine for his efforts. Schrödinger was very interested in a paper by Delbrück regarding the effects of ionizing radiation on genes, focusing on the nature of genetic mutation. Simply put, this paper presents the physics of genetic mutation. Besides mutation, another curious phenomenon is the level of fidelity in inherited genes (Al-Khalili and McFadden, 2015). Genes are known to be inherited with an extremely high degree of fidelity – approximately with a less than one error in a billion genes inherited. Explaining this mechanism with regular physics is not possible since conventional laws of physics involve broad-scale averaging. Such averaging will result in various errors in the inherited genes during the transfer. Additionally the regular laws of physics would not work if we are dealing with very small things like genes. This is why physicists such as Schrödinger and Delbrück were approaching biology using the lenses of quantum physics. So much so that Schrödinger dreamed-up his aperiodic crystal structure just for this purpose. In his view genes are constituted by this aperiodic crystal structure are:

1. More complicated than regular organic molecules (non-living) where each molecular behave in a strangely individual way – a 'quantum way'.
2. Capable of making quantum jumps within these crystals causing genetic mutation.

Before we delve into the details of what are quantum jumps and how they relate to genetic mutation, let us look at some elements involving inheritance. The process of inheriting genes could be simplified into two critical steps – crossover and mutation. Crossover is when genes from each parent organism are combined to become one. These genes exist as part of a larger structure called the chromosome. Half of the chromosome from each parent is combined to become a single chromosome in the child. This process repeats when the mates become parents themselves producing an offspring. For instance, human beings have 46 chromosomes where 23 are inherited from each parent. This applies for most organisms which reproduce sexually. Besides sexual reproduction, some organisms (mostly microbes) reproduce asexually needing only one parent. In such cases most of the genetics of the child would originate from the single parent, perhaps with slight mutations. In Figure 3.1 and 3.2, the chromosomes are coded by the letters; G, C, A and T. These letters represent the basic proteins that build genes – guanine, cytosine, adenine and thymine.

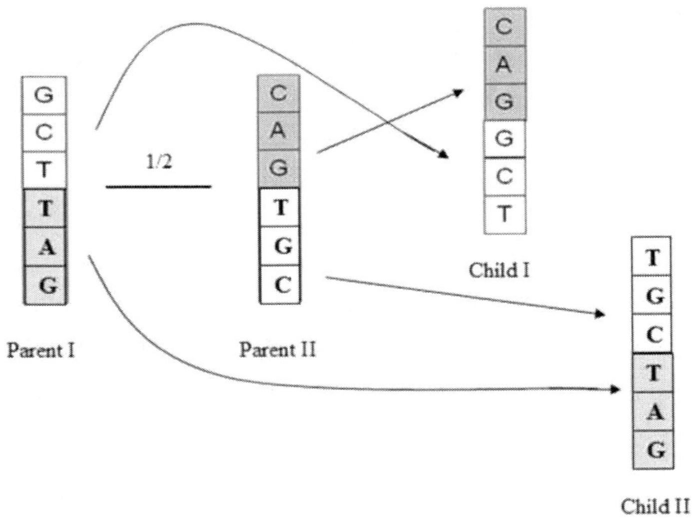

Figure 3.1. Genetic Crossover.

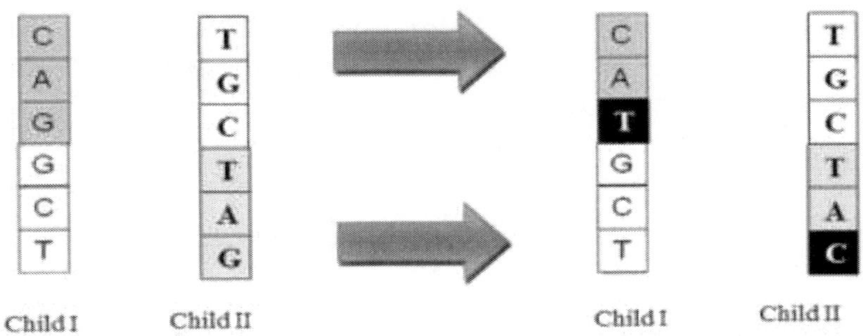

Figure 3.2. Genetic Mutation.

The gene codes in the chromosome structure are finite. So as reproduction continues, it is highly possible that at some point similar chromosomes would occur again in the offspring. This means that the original parents would somehow make a comeback in one of the future generations. Besides, having the same 'gene pool' may not be a good thing since the organism would not be able to develop new characteristics for adapting to the environment (e.g., sharper claws, faster running or smarter brains). This is where the second component; mutation comes in. Mutation slightly modifies the genetic codes during inheritance and during the organism's lifetime. It provides the organism with individuality as well as adaptation capability (notice the codes marked in black in Figure 3.2). This is why Schrödinger was amazed with Delbrück's article which tries to attribute mutation to radiation physics.

Coming back to the notion of quantum jumps and the role it plays in genetic mutation. Quantum physics was conceived in early 20th century before which classical physics was employed to explain how stuff works and other phenomena we happen to stumble upon. The core idea of classical physics is that properties like heat, electrical current and magnetism are continuous. These properties extend to the surroundings like long invisible threads which cannot be broken down to smaller entities. Besides being very effective in predicting real stuff, the continuity idea also went very well with the mathematics of the time – making it very easy to perform calculations which were then nicely associated to physical things. This was all until quantum physics came about. Quantum physics tackles physical problems occurring in very small scales. In such scales, the classical physics approaches almost always fail in explaining what's going on. Quantum physics disposes of the continuity outlook and says that everything is rather finite or discontinuous. This means that heat, electrical current as well as magnetism are finite. This translates to energy as being finite.

For instance, say you want to determine the heat transferred from some plate to the environment using classical physics. One could go on and calculate the corresponding heat transferred at any reasonable temperature. In quantum physics; such a calculation could not be performed – simply because the heat is transferred in finite 'packets' – which means that heat will only be transferred at specific temperatures (not at any reasonable temperature like the classical physics setting). The plate, therefore, transitions from one energy state into another as it releases the energy packets. This transition is also known as a 'quantum jump' or 'quantum leap'. The experimental proof of this idea called the photoelectric effect won Albert Einstein the 1921 Nobel Prize in Physics. Schrödinger stated that not only that genetic mutation may be caused by quantum jumps, but also that the potential of life to progress towards higher order and organization (towards lower entropy) springs from the quantum mechanisms operating at the core of its being. He calls this principle 'order from disorder'.

MEMORIES, GENES AND MAD COW DISEASE

As with the DNA, many biologist, psychiatrists and brain researchers have wondered how the mind works – especially the way it stores and retrieves information contained as memories. We now know that there are two types of storage mechanisms our brains use – short term and long-term memory. Short-term memory is a very simple form of memory employed by the brain to perform menial and simple tasks such as; driving, running, etc. Long-term memory on the other hand is used to perform complex activities – this type of memory is also sometimes called explicit memory. The part of the brain that carries out functions for long-term memory is the hippocampus. The same basic memory mechanisms are prevalent among many other organisms.

Among the most renowned scientist working on long-term memory is one Eric Kandel. Kandel was interested to know how memory forms, the way information comes into the hippocampus and how it is stored for long periods. The search for answers to these complex questions is daunting at the very least. To simplify the problem, Kandel took a different approach in his research. Instead of studying the complex brains of complicated animals, he turned his attention to the simple brain of the sea slug, *Aplysia*

(McNamee, E. and Wilson, J., 2013). Besides similarity in the basic memory mechanisms, the structure of human nerve cells (or neurons) is very close to those of snails.

Brains of most organisms function due to the communication between neurons – like a wired electrical circuit. Neurons could be thought of as the wires where an electrical signal is transmitted across the circuit from one wire to another. Besides electrical signals, the brain also uses chemical signals to communicate among neurons – or to communicate with other parts of the body such as muscles, tissue, etc. Each neuron is connected to one another by a bridge-like structure called 'synapse'. These synapses are said to be plastic; which means that they are modified when the brain learns new information. Using our electrical circuit analogy; the connections between the wires are altered during learning.

Having only 20 thousand neurons compared to a human brain (consisting of approximately 100 billion neurons), Aplysia grows up to a foot long. The large neuron size of this snail makes them highly visible – so much so that some nerve cells could be easily observed with the naked eye. Since they have very little neurons, researchers could easily record the activity of reactions in a single neuron. To top that, Aplysia also has learning capability – hence it could physically respond to various environmental factors. Such attractive properties make this sea slug a very desirable candidate for Kandel's brain research. Kandel found that the strength of communication between neurons (synaptic strength) change when the brain learns. When it comes to short-term memory this change in strength is only temporary. When the brain attains long-term memory, things get a little more interesting. Long-term memory is usually converted from short–term memory. According to Kandel's experiments; when long-term memory is formed, the brain does not only change its synaptic strength but rather alters the neural circuitry of the brain by growing new synapses. This change in neural connections is initiated by the synthesis of certain types of proteins. Let's go through these ideas in sequence (Figure 3.3). First the brain learns some stuff it labels as short-term memory. At some point the brain realizes this memory is essential and perhaps is required for understanding other more complex ideas. Now it decides to convert it into long-term memory. It then creates chemicals at the synapses. These chemicals act as messengers starting a series of molecular events (some reactions) which activates the genes in the neighboring neuron to produce the necessary proteins for growing synapses – altering the neural circuitry. The mechanism discovered by Kandel unravels the intricacies of how the brain forms long-term memory. This mechanism transcends the notion of mere electrical and chemical signals and probes further at a genetic level.

One may wonder why chemicals are needed to activate genes in the neighboring neurons. Why not simply produce the required protein and immediately grow new synapses? The thing is, proteins in our bodies are produced using instructions taken from the DNA. The DNA throughout the cells in our body are identical. However, different cells use various segments of the instructions and not the whole DNA code. For instance, cells in our eyes use instructions from a certain set of genes while skin cells use instructions from a different set. This way each cell differ according to their function – so they don't overlap causing wrong cells growing at wrong places in the body. The reading of information from the gene is called 'gene expression'. To put it more accurately, long-term memory is created by changing the gene expression at the neighboring neurons. The genes are expressed to create proteins that build new synapses in the brain. So for short-term memory, the brain uses existing neurons. But for long-term memory it uses genetic instructions to build new synapses which change the wiring of the brain. The strange thing is that the same synaptic-building agent for long-term memory known as prions could also cause mad cow disease in humans and cattle alike. Let us first explore what are prions and how they're involved in some types of diseases.

Various cells carry out their respective functions by building proteins using instructions from the DNA code. These proteins then do most of the cellular footwork. To optimize on a single protein's functionality, 'protein folding' occurs where specific types of folds could result in specific protein functions. This way, a single type of protein could be used for a series of functions by simply modifying the way it folds. Prions are proteins as well. But unlike regular proteins, they misfold and propagate throughout the organism by misfolding other regular proteins. This causes most regular proteins to lose functionality (transforming into prions) leading the organism to exhibit symptoms similar to a viral infection. *Prion* is derived from the words *pr*otein and infect*ion*. Prions are believed to cause Mad Cow Disease (bovine spongiform encephalopathy) in cattle and Creutzfeldt–Jakob Disease (CJD) in humans.

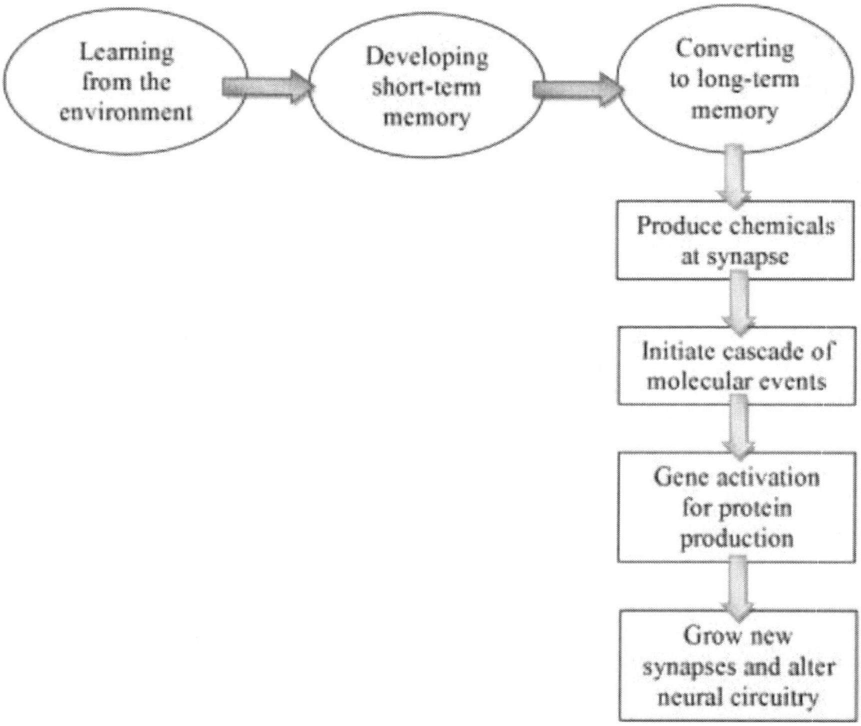

Figure 3.3. Developing Long-term Memory.

Besides CJD, they are also known to cause Kuru or ‘laughing sickness’, a neurological disorder found among tribes in Papua New Guinea. The disease is caused by prions found in humans who then transmit it to other humans. Like other prion diseases, Kuru is incurable and its onset leads to inevitable death. Reaching endemic levels among tribes in Papua New Guinea, Kuru is transmitted via funerary cannibalism – the traditional tribal practice of consuming the corpses of diseased relatives. Although there are different versions of prion diseases, they often have a lot in common. For instance prion diseases are often neurological. Once infected, they multiply and spread in the brain where they eventually start killing neurons resulting in rapid cognitive decline (Pocchiari et al., 2004). Although usually transmitted via cattle or humans, some prion diseases could occur genetically. Some people with certain inherited genes could be more predisposed to prion disease. These genes cause the production of misfolded protein forming prions (Bechtel, K. and Geschwind, 2013). Nevertheless the weirdest form of contracting prion disease is spontaneous. This happens by random chance where one protein suddenly misfolds leading to the misfolding of other proteins causing the onset of the disease. The exact cause of this type of prion disease is unknown. People with no prior exposure to prions and have no genetic predisposition get the disease out of the blue (Appleby and Lyketsos, 2011).

Coming back to long-term memory; Kandel found that new proteins are synthesized to grow new neurons for storing long-term information. But how is this information retained over lengthy periods? This is what Kandel and his team at Columbia University Medical Center set out to answer next. Continuing his experiments on Aplysia sea slugs, Kandel found that long-term memory is maintained with the help of a type of prion. Unlike the disease causing prions, these prions harmlessly contribute to memory storage. This prion called CPEB3 was found by Kandel to be present in mice (CUMC, 2015). Using mice they proceeded to test the truth of their hypothesis; that long-term memory is created with the help of prions. Making mice navigate through mazes stimulates the animals to make long-term memory. The researchers then turned off the gene that produces the prion CPEB3 and observed the mice navigate. They found that two weeks after the memory was made it disappeared – only those made with the prion-producing CPEB3 gene were retained. Prions generally come in two forms; soluble and aggregate forms. During the generation of long-term memory, soluble prions are converted into aggregate prions. These aggregate prions signal the initiation of protein synthesis necessary for maintaining memory over long periods. The presence of these aggregates ensures that long-term memory is continuously maintained. As Kandel says; “This ongoing maintenance is crucial. It’s how you remember, for example, your first love for the rest of your life. There are probably other regulatory components involved. Long-term memory is a complicated process, so I doubt this is the only important factor.”

Nevertheless these findings significantly change our perspectives on how we think of medicine and biological life. Whenever we think of disease and sicknesses – we immediately relate it to enemy pathogens like viruses. Or invasive organisms such as; bacteria, fungi and other parasites. In light of the findings of Kandel, we realize how the very thing (prions) which causes incurable diseases is the very thing that performs the fundamental function of long-term memory. It is possible that many of the pathogens and parasites that invade us do perform other functions that we have not yet discovered. Perhaps some of these functions are crucial to our on–going existence making them only malignant in certain forms. Similarly only certain proteins introduced into the body misfold and become threatening. On the other hand, the body’s own prions remain functional and is used to produce long-term memory. The situation is not unlike the development of cancer or tumor cells. Such cells multiply when one dysfunctional cell continues to propagate by making other cells which equally malfunction, making a growing group of malignant cells. Only dysfunctional cells become tumorous, other functional cells like functional prions are beneficial to the organism. A more complete understanding of the role played by pathogens and other disease-causing agents would definitely progress medical science towards treating various illnesses besides improving our knowledge regarding our place in nature.

LIVING COMPUTERS

From the works of Watson and Crick, it was clear that the DNA was a code script referred to for the production of proteins which carry out cellular tasks. The DNA contains the essential programming required to keep organisms alive. By the time of the discovery of the DNA, ideas from biology and computers were beginning to blend. Today this field is known as systems biology, where the cells are considered as little computers capable of producing other computers. Such cells operate using DNA codes running on some kind of a 'genetic program'. These studies were further spurred by the Human Genome Project (HGP) (An Overview of the HGP, 2016). The HGP was an international research program targeted towards the complete mapping and understanding of all human genes (genome). This way we could have a complete blueprint of the human genome. Discovering about 25, 000 human genes, the HGP unraveled three aspects of the human genome; (1) mapping the location of genes in major sections the chromosome, (2) producing maps that link inherited characteristics among genes which could be tracked over generations and (3) determining the sequence of bases in the human chromosome. Thanks to the HGP we now know detail information about the structure, function and organization of human genes (which directly influences the development of every human being). The HGP merely provides us with information about the human genome. Beneficial outcomes depend on how we interpret and utilize the information while dealing with social impacts such as; legal and ethical issues arising from the implementation of this new knowledge.

The advancement of computer technology significantly sped up the rate at which we acquired biological knowledge. The most basic building blocks of computers are electronic components known as transistors. Transistors are devices that act as switches in electronic circuits (they also act as amplifiers but for our purpose let's only think of them as switches). A typical transistor has three pins: base, collector and emitter. A picture of a real transistor is shown Figure 3.4 (left) while Figure 3.4 (right) depicts its symbolic representation:

Similar to other switches, transistors allow us to turn circuits on and off. The difference is that with transistors you don't really need to physically flip the switch, the base voltage does it all. As seen in Figure 3.4, electricity flows from the base to the junction either going to the collector path or the emitter path. Depending on the type, transistors come with a pre-specified voltage threshold (voltage is the potential for current to flow). Say we get a transistor with a 0.5V base voltage specification. Then if we supply electricity at the base above 0.5 V, then the current flows through the emitter path. If the electricity supply at the base is less than 0.5 V, current then flows to the collector. Therefore, the voltage at which we supply electricity at the base influences the path at which electricity flows. Let's say the emitter path supplies electricity to some electrical appliance; say the light to your room while the collector path discharges additional electricity by grounding it. By controlling the voltage of electricity at the base of the transistor we could switch your room lights on or off. Imagine if you have more transistors, say hundreds of them. Then by controlling the voltage supply (or more accurately by programming their voltage supply) we could control the way they behave. So much so that we could program them to perform math operations, compute solutions, control car engines and even make phone calls. Besides computers, transistors are used in your car's engine electronic control unit (ECU), chips in your mobile devices (smart phones, tablets, etc.) as well as most other devices which require some form of logical operation.

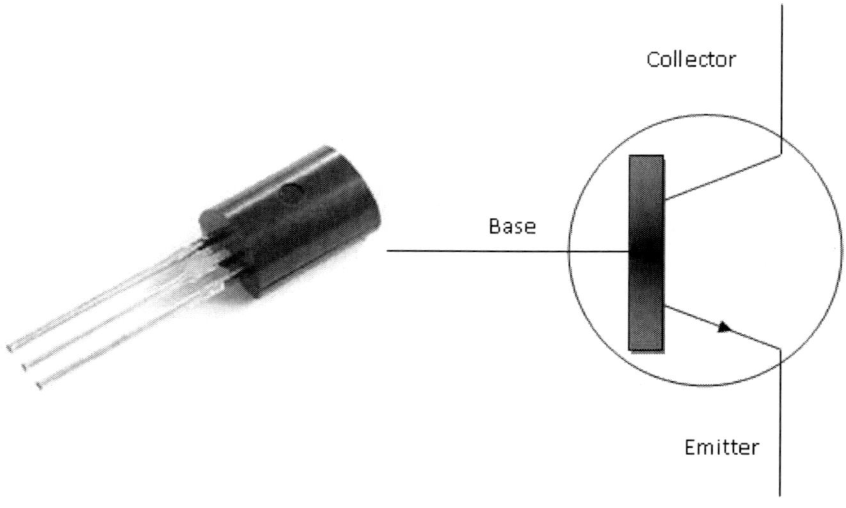

Figure 3.4. Transistor.

Recently, two research groups built a transistor made out of living cells (not with silicon or metal oxide-based materials like electronic transistors). Similar to their electronic counterparts, biological transistors act as switches used to perform logical operations. Drew Endy at Stanford University, created a type of biological transistor that controls the path at which the enzyme RNA polymerase moves (Heaven, 2013; Bonnet et al., 2013). RNA polymerases are intermediate enzymes functioning as copiers and carriers of information from the DNA. DNAs serve as sort of library to be referred to. Information on the DNA cannot be directly removed or modified. Therefore, the RNA polymerases copy and carry information to be used to make proteins. This process of copying and carrying is called 'gene transcription'. Gene transcription is analogous to parsing done by computers – where computers read and execute programmed codes. The part where the RNA polymerase enzymes come into play is shown in Figure 1.4 in Chapter 1. Main types of RNA polymerase enzymes are; messenger RNA (mRNA), transfer RNA (tRNA), ribosomal RNA (rRNA), micro RNA and catalytic RNA (Ribozyme). The biological transistor developed by Drew Andy and his team channels the flow of RNA polymerase the same way electronic transistors channel current in circuits. Such biological transistors could be used as fundamental components to build computing gadgets – much like electronic transistors. Timothy Lu and his research team independently built a similar gadget in Massachusetts Institute of Technology (MIT). Such 'bio-transistors' enables us to build more complex cellular machinery and instruments for a wide range of applications – especially for scenarios dealing with microscopic scales (or smaller). In such situations, the usage of electronic and other physical devices would deem inefficient and often not even functional. This is where the advancement of cellular machinery comes in. As previously mentioned, transistors are used to build other forms of computing devices besides computers. The same goes for bio-transistors, they could be used to build sensors, control systems and a wide variety of instruments – which could not be developed using regular electronic transistors. Drew Endy at Stanford University pushes the limits of our imagination by suggesting that in the future, such biological tools could give rise to the construction of living or organic bridges and buildings. Such structures would be made possible by directing the growth of natural materials achieved by manipulating its internal cellular circuitry. In medical applications, he speculates that someday we'll be able to send 'tiny submarines' with DNA computers to navigate through our bodies.

To make such things happen, focusing on physical hardware such as transistors and other gadgets alone is not sufficient. We would also need to master the software aspect of these bio-machines. Let's digress back to electronic circuits for a bit. Computer code and software have become commonplace today's world. They affect our lives in many ways changing the way we access information and communicate. Software directs the operations of computing devices. Thinking in terms of transistors; software determines which path the current must flow in a group of transistors which build a circuit. The connection of many such circuits make-up the electronic device. The software provides instructions on how the device should work but not how to design circuits. Engineers that create electronic circuits use a different language to describe their designs to other engineers. This is not software as we understand it, rather this is a language employed like a schematic or a blueprint to describe the inner-workings of specific circuit designs (called Hardware Description Language (HDL)). The standard HDL widely employed for this purpose is called Verilog. Using Verilog as a base, a new language for programming cells is being developed by Chris Voigt and his team. They call this HDL language for cells – 'Cello' (Coghlan, 2016; Nielsen, 2016). Working at MIT they are trying to program cells the same way we program computers. As Voigt says, 'We take the same approach as for designing an electronic chip. Every step in the process is the same – it's just that instead of mapping the circuit to silicon, it's mapped to DNA.' Manipulating genetic parts could be extremely tedious and describing this process on a blueprint similarly difficult. Cello (similar to Verilog) uses symbolic language to describe in short-hand the function of a DNA circuit. This way they obtained a sort of 'wiring diagram' or blueprint of DNA circuits. This diagram is then inserted into a machine that manufactures a DNA strand according to the blueprint. This DNA, engineered for a specific function is inserted into a microorganism. Voigt's team designed 60 such circuits and 45 of them worked exactly on the first try. Using this approach the team successfully manufactured one of the largest biological circuit ever built consisting of 7 logic gates. The circuits contained strands of DNA spanning 12,000 units in length. Using technologies like Cello, biologists could now make cells do stuff at a faster pace and not spend much time figuring out how to manipulate genes. Hoping that other scientists would use Cello to develop their own microbe, Voigt's team published details of their tool and approach online.

Speeding the process of genetic manipulation has many potential benefits which impact various areas of human life. At the same time there are potential negative effects which need to be addressed – we shall come to that later. These advances significantly improve the manufacturing of certain drugs and chemicals since not all such substances could be readily produced using conventional methods. Conventional here means manufacture by performing a series of chemical reactions. Excellent examples are the production of insulin for diabetic patients and the development of the Hepatitis B vaccine. These substances are not produced using chemical reactions alone, they require genetic technology and bio-engineering.

Genetically Engineering Medicine

Insulin is a hormone produced by cells in the pancreas for regulating glucose levels in the body. Glucose in the blood stream is converted from food and is used as an energy source by cells. Insulin acts as a 'transporter' bringing glucose from the bloodstream to cells. So when you consume food, insulin levels in the blood stream increases to transport the produced glucose to the cells. Once the transport is complete, the blood glucose levels return to normal in a matter of hours. If the pancreatic cells do not secrete insulin, then glucose builds-up in the blood since it cannot get to the cells. Starved of glucose, the cells are not able to generate energy to function properly. This is why diabetic patients often suffer from constant infections, fatigue, numbness, blurred eyesight and slow healing of injuries (Insulin, 2008; Clark and Russell, 2000).Diabetic patients are unable to produce the required amounts of insulin. There are two forms of diabetes; Type I and Type II. Type I diabetes occurs at a very early age and is usually the most severe kind. Patients suffering from Type I diabetes have pancreases that do not produce insulin. To supplement this deficit, they inject themselves with insulin three or four times a day. The insulin dosage is controlled based on their blood sugar level measured using a glucose meter. On the other hand, patients with Type II diabetes can produce insulin but not as much as the body requires (or their body is not responsive to insulin). This form of diabetes usually occurs in people who are older or suffering from obesity. These patients usually take insulin injections once a day.

Since the early nineteen hundreds, scientists have been gradually improving the development of insulin. In the early days, insulin was extracted from the pancreas of pigs and cattle; to be used as a supplement for diabetic patients. This is possible because the chemical structure of insulin produced by these animals is similar to the ones produced by humans. Large-scale insulin synthesis soon became more efficient with the advent of genetic engineering. Once biologists decoded the chemical structure of insulin, they quickly began to search for the gene responsible for producing insulin. By the late seventies, scientists were beginning to insert insulin genes from rats into a bacteria. This experiment was a success resulting in an insulin-producing bacteria – this way insulin could be manufactured in high volumes circumventing the ethical issues as well as other production problems related to the dependence on cattle and pigs as insulin sources. The method was so effective that in 1982 a company called Eli Lilly emerged producing insulin on a massive-scale. Eli Lilly became the first corporation to produce a genetically engineered product. Today yeast and a very common bacterium called Escherichia Coli (famously known as *E.Coli*) are used to produce insulin using gene splicing.

Another application of genetic engineering is the production of the Hepatitis B (Hep-B) vaccine. Hepatitis B Virus (HBV) causes a Hepatitis

B liver disease. Similar to regular vaccines, the Hep-B vaccine contains bits of the HBV RNA – allowing the immune system to recognize and develop defenses against HBV. Unfortunately the HBV vaccination process is risky. This is because sometimes the Hep-B vaccine contains full viruses that go into healthy humans causing them to contract the full fledge Hep-B liver disease, instead of strengthening their defenses against the disease. And this happens despite all the processing and purification done during vaccine production. In addition to this, the HBV could only be obtained from infected patients – presenting another setback. Similar to insulin, today Hep-B vaccine is produced by inserting a gene in the DNA (which controls the production of HBV fragments) into the yeast, *Saccharomyces cerevisiae*. This yeast then produces fragments of HBV which is purified and made into a vaccine. Such vaccines are called recombinant vaccines. Thanks to genetic engineering, Hep-B vaccine production now has a near zero risk of virus contamination.

The medical breakthroughs in the development of the vaccine against the Human Papilloma Virus (HPV) is another great example of the application of genetic engineering in medicine. HPV is the name of over a hundred different types of viruses affecting mostly women. These viruses propagate via sexual contact making it the most common sexually transmitted disease in the U.S. In most cases, the patient's own body's defenses could fight off the virus (Gardasil (HPV), 2006). However, there are cases when upon infection the virus could cause cervical cancer – or create abnormal lining of the cervix progressing towards cancer as well. Most cases of cervical cancer are caused by HPV infections making it the second most common cancer among women – first being breast cancer. Similar to Hep-B, HPV vaccines are produced by inserting gene codes from specific viral proteins into yeast. The microbe then produced these proteins. Once purified they become the HPV vaccine known today by its commercial name, Gardasil. As with Hep-B, there are no live viruses in the vaccine which eliminates the risk of getting HPV via vaccination.

Genetic engineering has helped resolve many problems arising in medicine. But these applications barely scratch the surface of the full potential of genetic engineering. Besides mass manufacturing medicine and creating safer vaccines, genetic engineering could be employed to identify and eliminate disease–causing genes (e.g., Alzheimer, Parkinson and Cancer genes). It could also revolutionize the type of medical tests we do on patients. Rather than testing them physically, a quick gene-check could provide a bit of information regarding potential illnesses and other complications. A potential application of genetic engineering is to modify the patient's genes to equip the immune system to battle current and potential infections. Many of these applications are still in their infancies and haven't grown from their research phases. Nevertheless, genetic engineering is seen to be very effective and is continuing to influence the path of modern medicine and healthcare.

ARTIFICIAL LIFE

The American scientist and entrepreneur, John Craig Venter is well-known for creating the first artificial life form on earth – a type of bacteria christened, *Mycoplasma laboratorium.* The controversial work involved more than 20 scientists over a 10-year period taking up a budget of 40 million USD (Sample, 2010). Venter's team built the genetic make-up of a bacterium from scratch using computers. They then inserted the synthesized DNA into a host cell. The resulting bacteria functions according to the instructions detailed in its engineered DNA. This way the first living organism was synthesized in a lab. In the past, Craig Venter and his team have experience transferring genetic materials from one bacterium into another. This was achieved with a technique known as 'recombinant genetics' – where a segment of DNA code from another species is combined with a host organism's original DNA. This technique has been most commonly used for vaccine development (mentioned in the previous section).

Craig Venter looks at his work as a software-developing method for programming cells. In his own words – 'We've now been able to take our synthetic chromosome and transplant it into a recipient cell – a different organism. As soon as this new software goes into the cell, the cell reads [it] and converts into the species specified in that genetic code. The new bacteria replicated over a billion times, producing copies that contained and were controlled by the constructed, synthetic DNA. This is the first time any synthetic DNA has been in complete control of a cell' (Gill, 2010). This sort of genetic research is aimed to solve major global issues such as global warming, pollution and the energy crisis – for instance; engineering microbes that consume carbon dioxide (or pollutants) from air or churn out biofuels. Venter's team has attracted a fair amount of criticism from various groups. Some groups have raised concerns examining issues on how the development of artificial life would impact us on a social level. Some of these concerns involve the usage of synthetic organisms in the development of bioweapons. The potential ecological harm from the introduction of synthetic organisms into the environment also raises red flags. The level and even the nature of the ecological backfire may be difficult to gauge in these scenarios. Let's not forget the threat of disease spread. Could these artificial life forms become some kind of infectious superbug capable of starting an epidemic? Perhaps, but threats aside, Venter's work also gives us a chance to peek into the building blocks of life itself – seeing what it takes to make a cell tick.

In 2010, his team set out to determine the minimum number of genes required for life to exist. This means they would have to perform experiments that trim off genes from bacteria without killing it. Six year later while performing these experiments, they discovered 'quasi-essential genes' a new addition to essential and non-essential genes. Essential genes are those that are critical for the organism's survival. These genes are closely related to the organism's environment. Non-essential genes on the other hand are not as critical. But they are built-in redundancies used as spare parts for DNA maintenance (Fang et al., 2005). The recently discovered quasi-essential genes do not directly jeopardize the organism's survival but significantly affects its growth (Hutchison et al., 2016). These experiments are opening up new directions in our current knowledge of the DNA.

Gene Editing

As our knowledge of the genome broadens, geneticists are developing better, more efficient tools for altering genes. The most advanced and simplest technique for gene editing is CRISPR (pronounced ‘crisper’). CRISPR stands for Clustered Regularly Interspaced Short Palindromic Repeats (it’s a mouthful and a tough one to remember). Before delving deeper into what is CRISPR, let’s explore a little about gene editing. The basic idea of gene editing is to be able add, delete, replace or modify genes at specific locations in the genome of a cell. Although it sounds simple, precise editing of particular genes can be tricky. What you would really need is a ‘molecular scissors’ which does all the high-precision genome slicing. Before CRISPR, older gene editing methods suffered from various setbacks making precise gene slicing difficult. Although many of these methods worked well on simpler organisms, they failed when it came to editing genes of more complex organisms. Due to these difficulties, multiple trials were required for successful editing – this of course translates to escalating costs. Since the discovery of CRISPR in 2015, these issues are slowly becoming something of the past. This new technique is considerably cheaper, reliable, more accurate and by far the most easiest to use (Tompa, 2016). So how does it work? There are bacteria which have gene editing mechanisms similar to CRISPR in their cells. Being part of their immune system, these bacteria use the CRISPR-like mechanism to defend against invading pathogens (e.g., other bacteria or viruses). So when they are attacked by a virus, they respond by precisely snipping out part of the virus DNA injected into them. They then store the DNA segment. This way they could recognize the same virus and defend themselves if they are attacked again in the future. Cas9 enzyme acts as a ‘molecular scissors’ which does the precise snipping of the virus DNA code – this is why the CRISPR technique is also sometimes knows as ‘CRISPR-Cas9’. So biologists took this idea and used to edit the genes by introducing mutations at certain parts in the genome. First an auxiliary piece of RNA called the guide-RNA (gRNA) is designed. This gRNA functions as a guide which takes the Cas9 enzyme to the exact location in the DNA. Reaching there the Cas9 enzyme does what its best at – slices the DNA segment. This creates a void in the DNA which is soon noticed by the cell. The cell then starts to repair the lost DNA segment creating a mutation. Thanks to precise snipping using CRISPR, geneticists can now create very specific mutations and observe its effects on cells. Development of CRISPR and other similar approaches usher a new age in gene editing. Such powerful tools create new ways of approaching some of the long-standing problems in biology and medicine.

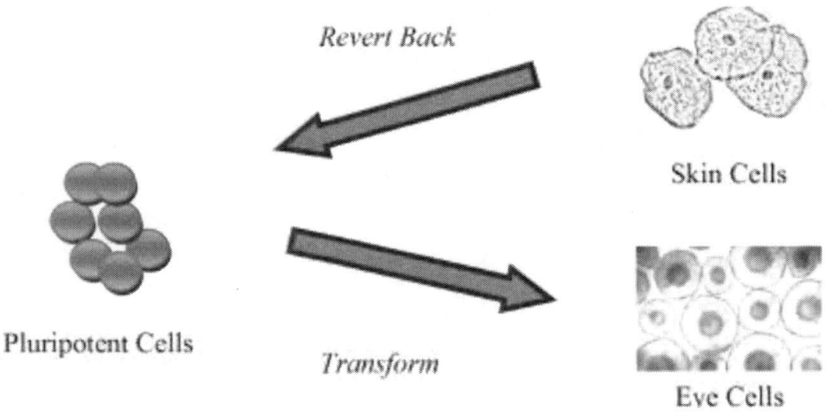

Figure 3.5. Yamanaka Approach.

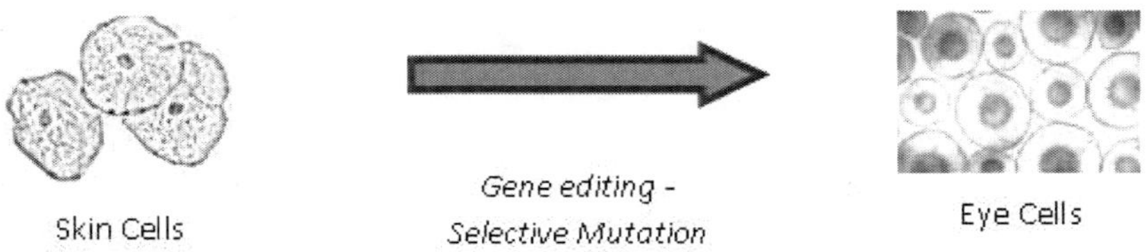

Figure 3.6. CRISPR Approach.

Among the most impactful fields in medicine is stem cell research. Stem cells also known as pluripotent stem cells are initial-stage cells that have the ability to differentiate into any type cell (e.g., hair, eye or skin cells). If we could control how stem cells differentiate, then we wouldn't have to rely on foreign organ transplants. An organ could be simply designed by using the patient's own stem cells. Such a technology would avoid many complications arising from organ transplants; from lengthy patient's waiting periods to difficulty in finding suitable donors. A Professor from Kyoto University, Shinya Yamanaka discovered in 2006 how to revert adult tissue cells back to their pluripotent origins (Figure 3.5). This won Yamanaka the Nobel Prize six years later (Kingery, 2016). Researchers, therefore, continued progress by making many pluripotent stem cells (creating a 'master key'). These pluripotent cells are then differentiated into the desired cells. Instead of creating many 'master key' pluripotent cells, using CRISPR researchers could make differentiated cells reprogram themselves. Meaning that they don't need to revert differentiated adult cells back to their pluripotent origins. Using CRIPR they just need to perform a bit of gene editing (via selective mutation) which turns on or off certain parts of the genome – this in effect modifies its functions changing the cell-type (Figure 3.6).

The advent of CRISPR spurred technological developments in various fields. Debates have been held on the potential use of CRISPR as a tool to battle deadly diseases such as chikungunya, dengue and recently, Zika – propagated by the mosquito species; *Aedes Aegypti* (Bogoch et al., 2016). Up till now, humans have used insecticides, bed netting and other types of screening techniques to filter out these mosquitoes from reaching our bodies. Unfortunately mosquitoes have evolved various types of resistances against pesticides and they seem to eventually find ways to penetrate our screens and feed on our blood. Zach Adelman, a virologist from Texas A&M University is working on a technology to wipe out all *Aedes Aegypti* mosquitoes (McKay, 2016). His idea is to perform gene editing on these mosquitoes causing them to produce only male offspring. Therefore, the next generation of offspring would soon run out of mates to reproduce resulting in mass extinction. Additionally only female mosquitoes of the species feed on humans and propagate diseases. Besides halting the mosquito population growth, an 'all-male' population would put a stopper on disease spread. This type of CRISPR approach is called 'gene-drive' – where selective mutation is performed to increase the chances for these genes to be carried into the coming generation. All this is good and well, but is wiping out an entire species for the sake of our survival a good move? One thing to consider is how these parasites are linked in the ecosystem. By understanding what role they play, we could then get a better idea on how to deter the disease vector. For instance, even these harmful (mostly to humans) mosquitoes may help pollinate plants, eliminate certain bacteria/fungus or support the survival of some other organism(s). As mentioned in Chapter 1, parasites often have very complex lifecycles. Meddling with their populations might result in extreme changes in the ecosystem which may cause us more harm than the parasite itself. Gaining a better understanding of these mosquitoes' lifecycles would be advisable before we go on eradicating an entire species. Besides gene-drive, there is another technology developed by Oxitec Ltd. This method inserts self-limiting genes which causes the mosquitoes to produce offspring that die or become sterile. This causes population decline but not total extinction – unlike gene-drive, the genes are not propagated to the subsequent generations since the immediate offspring either become sterile or die. Perhaps by using CRISPR we could reach sort of a middle ground between those two approaches where we could decline the *Aedes Aegypti* mosquito population significantly without driving it to extinction. It would be interesting if we could identify a way to control the gene-drive mechanism so that it only kicks-in after some number of generations. Meaning that at certain generations there will be more male mosquitoes than females. This would shrink the overall population just enough so that it's not harmful to humans.

CRISPR has also been used to shed some light on the emerging field called 'evo–devo'or evolutionary developmental biology. Evo-devo is a field that compares and analyzes biological developments among organisms (e.g., how limbs develop or how butterflies evolve color vision). The idea is to understand the ancestral connections between different life forms. Scientists in evo-devo are starting to use CRISPR to observe evolutionary changes in organisms. This is how they do it; first they remove certain gene segments using molecular scissors, Cas9. For instance; if they slice-off the genes which are supposed to be for the development of the organism's legs, then by selective mutation, the cell would replace the leg-genes with some other genes (the predecessor of leg-genes). Therefore, instead of developing legs, the organism will start growing what it used to have before it evolved legs. Using CRISPR, scientists could avoid speculating and pursuing theoretical debates on how organisms developed (to the form that we see today) and start observing these changes in laboratories (Nakamura et al., 2016). The most exciting gene editing experiment in evo-devo was conducted by Neil Shubin in the University of Chicago, Illinois, U.S (Callaway, 2016). Shubin wanted to know if fins and rays in fish were replaced by feet and digits in four-legged vertebrates during the course of evolution. This was inspired by the discovery of a 375 million year old fossil by Shubin's team in 2004. The ancient fossil was a type of fish called *Tiktaalik Rosseau* (Boisvert, 2005). Researchers describe this fish as evolution 'caught in the act' where fishes start evolving legs to dwell on land – becoming many modern land-dwelling animals we see today. Using CRISPR, Shubin removed the 'fin rays' genes called *hox13* from the genome of zebrafishes. These mutant zebrafishes did not grow full feet but rather some grew finger-like fins. These fins looked like bones that build fingers and toes in four-legged land-dwelling animals (tetrapods). Used like this, CRISPR could be used to guess DNA sequences of various prehistoric creatures – a 'Jurassic Park' scenario.

Directing Evolution

The CRISPR approach has propelled the notion of 'gene drive' to be seriously considered for implementation. As mentioned in the previous section, the central target of gene drive is to curb the spreading of diseases propagated by the mosquito, *Aedes Aegypti* (responsible for dengue and Zika infections). Usually the offspring of any organism have a 50–50 chance of inheriting any trait from either of its parents. With gene drive, traits can be 'driven' into subsequent generations at almost a 100 percent chance. For instance, if you want the *Aedes Aegypti* mosquito to have three wings instead of two, design a gene drive for three wings and incorporate it into the parent mosquito. Its offspring and subsequent generations will all be three-winged.

The main catalyst hoisting gene drive into reality is CRISPR. Using CRISPR, gene drives could be easily incorporated into the genome of almost any organism. This is how CRISPR comes into play: first one CRISPR-equipped gene is inserted into one of the parent (let's call this Parent 1). After mating, the genome of Parent 1 meets its counterpart; the genome from Parent 2. The CRIPSR-equipped gene from Parent 1 snips-off the corresponding gene at Parent 2's genome. Parent 2 then attempts to repair its genome accordingly – where it uses the coming genome from Parent 1 as a replacement. Existing now in both parents' genome, the CRISPR-equipped gene is driven to the offspring's genome. This drive continues when the offspring mates to produce the next generation. Figure 3.7 depicts this process.

Recently researchers have found that gene drive may not result in a 100 percent inheritance all the time. Lab experiments conducted on malarial mosquitoes (*Anopheles gambiae*) show that the mosquitos could develop resistance to gene drive (Hammond et al., 2015). Researchers in the US developed simulations to calculate the odds of gene drive not working (Unckless et al., 2016). Computer models become handy for cases like these – when experiments at large scales deem too difficult to conduct. Once you have performed gene drive on some organism, releasing into the environment means that whatever the implications, the process cannot be reversed. It may have unpredictable and possibly have grave impacts on the balance of the ecosystem. A useful concept that could come to our aid when facing these sorts of situations are natural experiments (discussed in Chapter 2). Biologists could keep a look out for organisms developing traits with near 100 percent chance of inheritance in coming generations. They could then observe what happens to these creatures in their natural setting – in addition to providing insight on what could cause 'natural gene drives'. On that note, it would be interesting if we could have some sort of simulation which analyzes what-if scenarios – a sort of disaster simulation. In engineering applications, concepts such as system risk and reliability simulations are common. Such simulations are carried out on aircraft components, chemical plants and power systems to identify high-risk factors. This is done to simulate the system's potential failures in the event of a malfunction. Such disaster simulations could lend a hand when investigating potential hiccups of gene drive while answering what-can-possibly-go-wrong questions that arise from it.

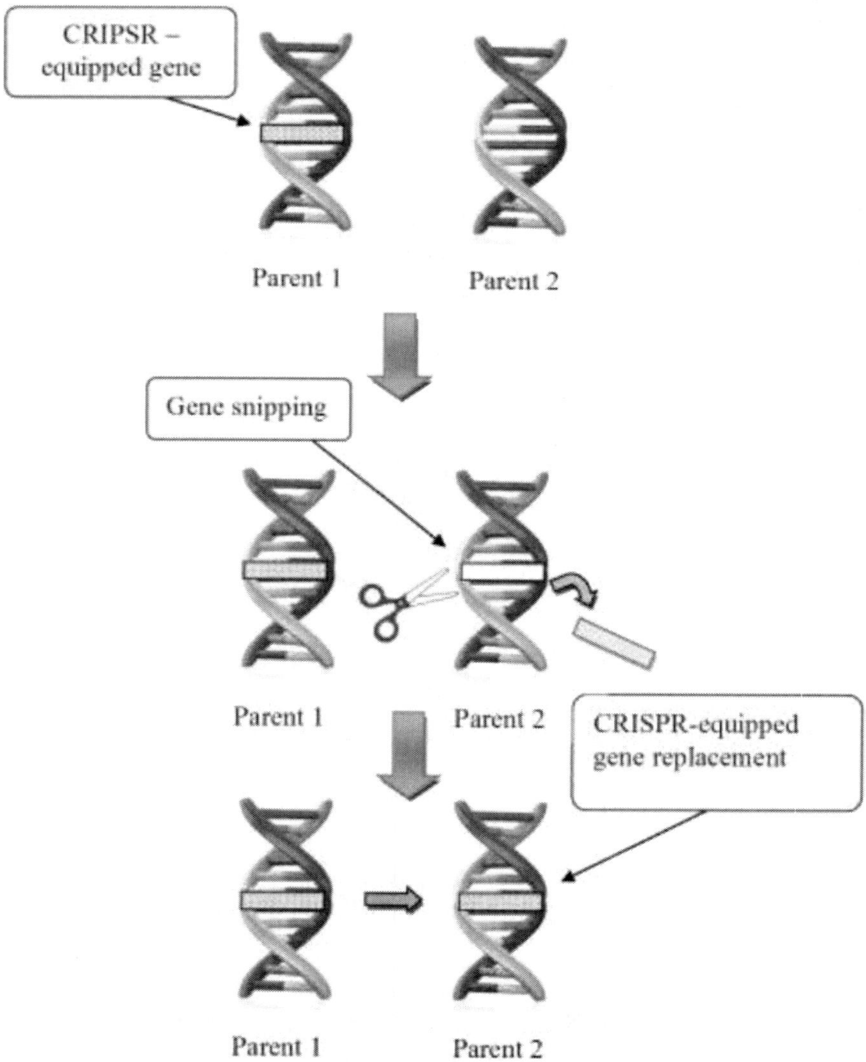

Figure 3.7. Gene Drive Mechanism.

Resistance to gene drive is directly related to the way DNAs heal themselves. The most extreme case of DNA healing could be observed in a bacterium called *Deinococcus radiodurans*. The second part of its name (*radiodurans*) translated from Latin means 'radiation surviving' – that's right, this species of bacteria have the ability of surviving high-levels of radiation without suffering from DNA damage. *D. radiodurans* was first seen during the canning of ground meat around 1956 (Cox and Battista, 2005). The bacteria was discovered when it was bombarded with high levels of radiation (gamma ray) to sterilize the meat inside from any existing bacteria. This is why canned food have very long shelf-lives. About 5 Gy of radiation would kill a human. *D. radiodurans* could take a load of 5000 Gy (Gy is short for Gray; the unit used to measure radiation levels). These bacteria are so resilient they could thrive in diverse environments – they have been observed in elephant dung and known to exist in granite on the dry valleys of the Antarctic (DeWeerdt, 2002). The secret behind the *D. radiodurans'* resilience is its ability to repair its DNA efficiently, even after sustaining extensive damage. These bacteria store many copies of its genome (four to ten backup duplicates). Besides, it also has very advanced DNA repairing mechanisms. At a glance it may seem as though *D. radiodurans* are built prepared to survive nuclear winters. The strange thing is that these bacteria can withstand high radiation levels which are never found naturally in any place on earth. One theory is that the bacteria evolved this trait as a response to dehydration. This is because severe dehydration damages the DNA similar damage caused by radiation. *D. radiodurans* could also survive in vacuum, extreme temperatures and highly acidic conditions. It's possible that the radiation-resistance is a by-product of its resistance to other harsh conditions.

Returning to gene drive and DNA self-healing; what we saw in *D. radiodurans* is what we call 'homologous repair'. Homologous repair is replacing a damaged or absent segment in the genome with another code which functions similarly (a spare part). This is how *D. radiodurans* repair their genome, by replacing the damaged segment with a similar code from its multiple backup DNAs. The inverse of homologous repair is 'non-homologous repair'. In this type of repair, the damaged segment is replaced with random code. This is the type of repair which cultivates resistance to gene drive. This is because once the parent gene code is snipped by the CRISPR enzyme, the appropriate replacement is expected to happen (homologous). When this does not happen, and a non-homologous repair (random) occurs, the CRISPR-equipped gene replacement won't occur properly (see Figure 3.7). Therefore, the gene-drive would not be passed down to subsequent generations. Robert Unckless, a geneticist form the University of Kansas puts it this way; 'If you can't get non-homologous end joining under control, resistance is inevitable. But resistance could take a while to spread, which means you might be able to achieve whatever goal you want to achieve.'

So random gene repair can deter gene drive. Population dynamics is another thing that can put a stop to gene drive. For instance, a population of organisms could simply not mate with the individual with gene drive. Even if gene drive works, evolution may react resulting in the organism developing some other type of similar trait. If we use gene drive to wipe out a species of mosquitoes (e.g., *Aedes Aegypti*) – it's possible the dengue and Zika pathogens may adapt to other mosquito species making them their new host. Even with powerful technologies like gene drive, manipulating the evolution of organisms is no small feat.

Onwards with Genetic Engineering

Genetic engineering encourages interdisciplinary research fusing biology, engineering and physics. It has come to known that the most efficient energy systems are biological. Over evolutionary timescales (millions of years), nature has gradually improved the design of photosynthesis in plants – yielding an almost hundred percent efficiency in energy conversion. Even our best energy systems, be it solar cells or combustion engines could not mimic such high efficiencies. Scientists have come to know that such biological processes in nature exploit the effects of quantum physics to achieve such feats. Biological processes employ quantum phenomena to fine-tune the efficiency and accuracy of their systems – these ideas would be discussed in greater detail in Chapter 5.

Researchers at MIT and Eni (an Italian energy company) developed an energy-efficient system using genetic-engineered viruses (Park et al., 2015). The research team encompassed of fourteen scientists including a genetic engineering expert, Angela Belcher and a specialist on applied quantum theory, Seth Lloyd. During photosynthesis, a photon which is a quantum particle of light is received by a chromosphore (molecule responsible for color). The chromosphore then produces an exciton, which is a quantum particle of electricity. Jumping from one chromosphere to another the exciton travels until it reaches the reaction center. Although the whole process is happening on a quantum scale, it's not so different from electricity traveling through a 'wire' – with chromospheres being the 'wire' and the excitons representing 'electricity'. At the reaction center, the exciton is harnessed to do biological stuff like cell metabolism, producing molecules and other things critical to support life (Chandler, 2015). Jumping from one chromosphere to another, the exciton needs to try each route randomly. This of course is tedious and would result in energy losses along the way making its transport inefficient. Therefore, the exciton uses quantum phenomena to expedite its travel, where it employs superposition to travel multiple routes simultaneously and select the best route. Or in quantum mechanics lingo, the exciton 'collapses' to its best route. During this process the exciton behaves more like a wave than a particle. The idea here is to create the conditions just right for this sort of quantum travel.

Belcher managed to engineer a virus which bonds to artificial chromosphores (a type of organic dye). The research team managed to experiment with various types of viruses to identify the best chromosphore for the most efficient exciton transport. In Lloyd's words – 'I had been talking about potential systems you could use to demonstrate this effect, and Angela said, "We're already making those." We came up with design principles to redesign how the virus is capturing light, and get it to this quantum regime." Thanks to genetic-engineered viruses with a dash of quantum mechanics; a potential method for designing cheap and efficient solar cells and other energy transfer devices have been discovered.

In the previous sections we discussed how the research teams of Timothy Lu and Drew Andy developed 'bio-transistors'. Thanks to these developments, researchers at University of California, San Diego (UCSD) genetically programmed the bacterium, *Salmonella enterica* to convey anti-tumor toxins to cancer cells in mice. The bacteria's genes were programmed by creating a simple genetic circuit (similar to the RNA polymerase-type circuit). The circuit was developed by Jeff Hasty, a biologist at UCSD, testing it initially on the bacterium, *E.Coli*. The circuit allows systematic toxin delivery and timely self-annihilation of the bacteria, rendering it harmless to the patient (Din et al., 2016; Azvolinsky, 2016). The procedure itself is still being researched and not proven for effective eradication of cancer cells in humans. Nevertheless, the mice experiments seem to show promise and most importantly it proves the potential strength of the genetic programming paradigm.

TRANSGENIC FOOD

We have seen some pretty exciting stuff when tech meets genetic engineering. But when genetic engineering gets mixed up with the food and medicine industry, that's when we get a little skeptical. Among the most controversial aspects of genetic engineering is the way it's used to modify food. We know this as genetically modified food – or in short GM food. Just like conventional food, GM food is grown in farms. The difference is that the DNA of the grown food (plants) is modified using recombinant DNA technique. Recombinant DNA is a technique that combines a foreign piece of DNA into a host. The foreign DNA usually comes from a completely different organism compared to the host. For instance, mice have been made to glow by transferring specific jellyfish genes into them (Okabe et al., 1997). Organisms like those mice are called 'transgenic' animals. Previously we discussed how insulin is mass-produced using the bacterium, *E. coli* – done by transferring human insulin genes into the bacterium. Similarly GM plants contain foreign genes, giving these plants advantages like: additional resistance to parasites or faster maturing rates. Unlike the CRISPR technology, the recombinant DNA approach is slightly older – since the days of genetically-engineered insulin. The Recombinant DNA approach uses a virus to act as a carrier (known as a vector) between the 'host' and the 'donor' organism. The donor provides the genes while the host is at the receiving end. In the case of the glowing mice engineered using jellyfish genes – the jellyfish is the donor organism and the mice are the host. Therefore, the vector (like a retrovirus) carries the donor's genes into the host organism and transcribes these genes into the host DNA. These engineered and grown plants become GM food. This way, beneficial genes are picked out from other foreign organisms and then inserted into the plant seeds before they are grown.

Such clear advantages are found in transgenic potatoes. Potato blight is a common fungus which ruins plantations worldwide. To defend against blight, farmers need to keep spraying pesticides up to 15 times a season. Another line of defense is to interbreed potatoes with their wild relatives – hoping that blight–resistant genes would cross-over to the grown potatoes. Unfortunately this process is slow and tedious with very low success rates – since the blight could evolve to overcome the resistance from these wild potatoes before the genes cross over. Some years back, scientists in the UK started running experiments on transgenic potatoes created by splicing genes from a South American relative and inserting them into Desiree potatoes (Jones et al., 2014). The idea is to create blight-resistant transgenic potatoes. Besides resistance, the shelf-life of fruits like bananas and tomatoes could be extended by genetic engineering (Elitzur et al., 2016). This way, suppliers could avoid using harmful food preservatives.

Fish farming has also been a target application for genetic engineering. During winter, Atlantic salmon grow very slowly. Recently a genetically modified variant of the salmon has been approved by the US Food and Drug Administration (FDA) for distribution. This GM salmon matures at a faster rate than its natural counterpart – making it suitable to be grown in tanks and other urban facilities. This approval marks the first time a genetically engineered animal has been approved for food anywhere in the world (Briggs, 2015). This fast-maturing salmon raises myriad ecological concerns – especially if accidentally released into the environment. Speedy maturity may upset the ecosystem's balance – the Atlantic salmon's fast maturity may result in a spike in predation, depleting the number of its natural prey at an advanced pace. This would create a sort of 'ecological shock' for other organisms connected to the salmon. There are many ways such sudden changes could impact the ecology.

Another type of GM crop which is thought to significantly affect the ecology is a corn variant called 'Bt corn'. The name comes from the spliced gene (Bt) taken from a soil bacteria called *Bacillus thuringiensis* which is inserted into the corn genome. Bt corns produce toxins targeted to exterminate certain insects – albeit harmless to humans. Bt corn plants have often been plagued by the European corn borer; a kind of caterpillar that bites into the leaves and stems of the corn plant. So when they prey on the Bt corn, the plant releases a toxin which attacks the digestive system of the caterpillar killing it in couple of days (Gassmann et al., 2014). Therefore, farming Bt corn requires cuts the usage of pesticides. Coming back to our ecological issue; in the nineties, Bt corn pollen was shown to harm monarch butterflies. Although the butterflies themselves don't consume corn but rather feed on milkweed plants which happen to grow in corn farms. Fortunately the level of monarch butterfly casualties from Bt toxins were deemed negligible. That said; the ecological concern is more complex than simply the reducing of numbers in a single species. A likely scenario could be; since Bt corn plants are commonplace, many other insects feeding on nearby plants (like the Monarch butterfly) develop resistances against the toxin. Besides Bt, these organisms could develop resistances to similar naturally-occurring Bt-like toxins (Badran et al., 2016). Consequently some species would increase radically thanks to this new found resistance; which in effect causes various new interactions to exist in the ecosystem. Over a period of time the genes developed by other organisms in response to the Bt corn plant move through the generations shifting and channeling evolution. Scientists call this phenomena 'gene flow'. Many ecologists worry that gene flow from these GM plants to various species in the ecosystem would trigger unforeseen and unprecedented long-term ecological impacts. As stated by a plant ecologist at the Ohio State University, Allison Snow – "We've let the cat out of the bag before we have real data, and there's no calling it back. Genes flow from crops to weeds all the time when pollen is transported by wind, bees, and other pollinators, even a very low probability event could occur when you're talking about thousands of acres planted with food crops" (Ackermann, 2002).

Nevertheless the many global benefits GM food has to offer never goes unnoticed. A good example is the GM rice famously known as 'golden rice' or its scientific name, *Oryza sativa*. The genome of golden rice is spliced with carrot genes to increase its content of provitamin A (provitamins are compounds which are converted to vitamins in the body). The idea here is to address vitamin A deficiency suffered by many poorer communities worldwide. An estimated 250 million children in preschool have vitamin A deficiency globally – with a staggering 250 000 to half a million children becoming blind every year from this deficiency (WHO, 2016).

Genetically-engineered plants and animals for food is not a new idea. For many years mankind has cross-bred organisms, transferring genes among species to ensure stable and desirable qualities in their food stock. The only difference with current technology is that gene transfer could be done more accurately and efficiently. Nevertheless many concerns remain to be addressed before embracing GM food products. Each food products would have to be individually tested to ensure low risk to humans in terms of: toxicity, antibiotic resistance and possibility of the existence of allergens (substances that could cause allergic reactions). Regulatory bodies like the FDA should continue designing test protocols for evaluating GM food prior to distribution (e.g., testing for viral contamination). Ideally, such approvals and testing should be incorporated into the legislation of every country and GM food labeling should also be made compulsory so that the public is well aware of the risks concerned.

Our Kind of Virus

Most of the time instructions in cells are executed in a specific sequence – the DNA in the nucleus is read, copied and transported to the ribosome by RNA (mRNA and tRNA). At the ribosome, amino acids are assembled to build protein molecules which perform specific cellular functions (DNA $\rightarrow$ RNA $\rightarrow$ Protein). The process where the DNA is copied to the RNA is called transcription. It has been discovered that some animal species could hack the operations of their cells deliberately by intercepting the flow of information during gene transcription. One species of squid called *Doryteuthis pealeii* has the ability to do exactly this (Alon et al., 2015). After transcription, the RNA is intercepted and recoded before it reaches the ribosome. This way the squid alters the shape (fold) of the protein directly affecting the protein's function. RNA recoding is believed to give the squid an evolutionary edge for adapting to rapid environmental changes – especially changes in temperature. Another animal that could naturally do this using a similar mechanism are fruit flies, editing about 3% of their mRNA (Luntz, 2015).

The animals we talked about earlier modify their cellular functions by hacking information at the RNA; they do not tamper with the DNA code itself. Nevertheless there are instances when such tinkering occur – known as reverse-transcription. During reverse-transcription; instead of copying, RNAs start inserting codes directly into the DNA. Reverse-transcription falls into the realm ruled by 'retroviruses'. Retroviruses are certain viruses that have the ability to recode the DNA of its host cell e.g., Hepatitis B and HIV viruses. Once a retrovirus infects a host cell, it then creates a DNA segment from its RNA using enzymes. This viral DNA is incorporated into the host cell's original DNA. Going unnoticed, the viral DNA is transcribed along with the original DNA in the host cell. Not knowing better, the host cell executes commands from the sinister viral code. Retroviruses like Hep B, HIV (human immunodeficiency virus) or rhinovirus (flu virus) come from external sources, spreading from other people or contamination from the environment.

Strange still viruses also exist internally within our bodies. These viruses are called endogenous retroviruses (ERV). ERVs are inherited into our DNA from past generations. Human DNA consists of a shocking 8% viral DNA – about 100, 000 pieces of ancient viral code. Since only 1.5% of the DNA are codes for human genes, we are genetically more virus than we are human (Wildschutte et al., 2016). Most of these ERVs have accumulated mutations over the past making them inactive.

A type of ERV called HERV-K has been linked to amyotrophic lateral sclerosis (ALS) disease (Li et al., 2015). Also known as Lou Gehrig's disease, ALS is a motor neuron disease that causes neurons which control muscle movement to die. Patients suffering from the disease gradually lose control of their muscles. Links between neurological disease and ERVs have been emerging since 2006. At the time a neurologist, Avindra Nath observed such links in patients suffering from both HIV and ALS. In one case, the patient's ALS symptoms improved when he took antiretroviral drugs for treating his HIV. Nath also discovered that both diseases sometimes occurred simultaneously. There have also been cases where doctors have found patients with viral infections but no sign of any known external vinfections. It was thought that such symptoms could only result from innate or endogenous viruses. Recently Nath's research group activated certain viral genes in mice using the CRISPR method. They found that these active genes caused the mice to develop ALS-like symptoms. So ERVs are harmful if activated, but why are they in us in the first place? And where do they come from? Scientists believe that ERVs become part of the DNA from viral infections in the distant past; millions of years ago. Originating from the outside environment, these retroviruses are said to infect the host organism, where they perform reverse-transcription. To prevent and ready itself for future encounters with similar ERVs, the host organism incorporates and maintains part of the retroviral code in its own DNA. This helps the host cell recognize similar viral infections.

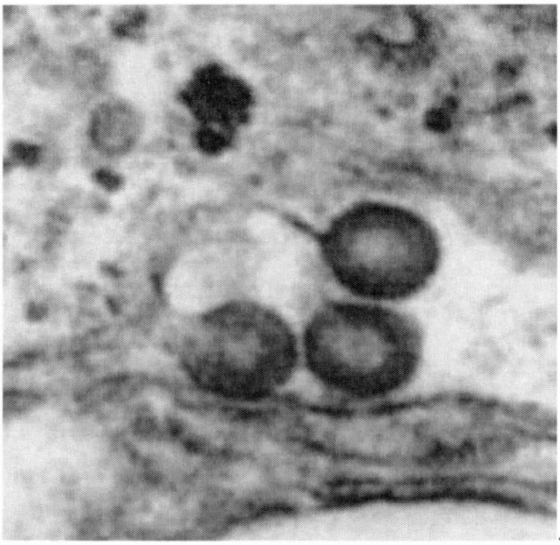

Figure 3.8. Electron micrograph of a placental tissue of a rhesus monkey taken in 1970 (Feldman, 1979).

A team at the University of Utah identified such a viral gene in the human DNA – which they called MER41 (Chuong et al., 2016). They said MER41 comes from viral infections dating back to about 45 to 60 million years ago. Similar to Nath's team, this group employed the CRISPR method to delete MER41 genes to see how cells respond to viral infections. The team proceeded by exposing a cell absent with MER41 gene to *vaccinia* virus (a viral strain related to small pox). It was observed that without MER41, the cells are more vulnerable to viral infections as compared to the unedited cells. These experiments show that ERVs are a vital component in the organism's immune system. Besides humans, researchers have found genes similar to MER41 in other mammals including bats and lemurs.

The most peculiar aspect of ERVs is their connection to our prenatal development. The placenta is an organ that connects the fetus to the uterine wall. This connection allows the fetus to; access its mother's nutrient, dispose its waste and regulate its body temperature. The most important role played by the placenta is the protection of the fetus from being attacked by the mother's own immune system. Since the fetus's DNA is not similar to its mother's but a combination of both parents, the mother's immune system may flag it as a foreign invader (or a parasite). This causes the mother's immune system to attack and eliminate the invader. Thanks to the placenta, this sort of thing doesn't happen since it acts as a barrier separating the two bodies. Evolved about 150 million years ago, the placenta appears in many other mammals (Henzy, 2014). It evolved to provide sustenance and protection for stable growth of the organism prior to birth. Retroviruses were first seen to be budding in placental tissue in the early seventies. Figure 3.8 shows a picture of retroviral particles in placental tissue of a rhesus monkey taken using an electron microscope:

The prevalent idea is that; at some point in evolution, mammals were infected by retroviruses. Incorporated into the DNA, the viral code plays an essential role for developing the placenta by creating a protein called *syncytin*. Syncytin is used to create the fusion layer between mother and fetus. Interestingly many viruses also use syncytin to fuse host cells together so they can jump from one cell to another efficiently. It happens that syncytin is not only present in humans but also other primates like chimpanzees and gorillas. All primates (including humans) were observed to have the similar genes that produce syncytin – probably due to our common ancestry.

In 2005, a French research team headed by the virologist Thierry Heidmann discovered that syncytins were not only unique to primates; they found some of it in mice (Dupressoir et al., 2005). Identifying the corresponding viral gene in mice, they set out to determine the importance of syncytin–producing viral genes in mammals. They proceeded by turning off the viral genes in mice embryos. They found that the embryos died in about 11 days – underscoring the essential role syncytins play during initial development of mammals. It was later found that though functioning the same way, the syncytin produced by mice and primates differ considerably. Syncytin produced by rabbits also differ from mice and primates (Heidmann et al., 2009). Curious still, the pika being a close relative to the rabbit does not share its syncytin type. Heidmann's team even found that many carnivorous mammals; dogs, cats and hyenas have a same type of syncytin – they labeled Car1. What virologists have discovered is that; infected by different viruses many animals mammal, rodent or feline have syncytin–producing genes. Though there are various types of syncytins, they all function the same way; to develop the placenta during reproduction. This is to say that infections from different viruses at different points in time resulted in the production of the same protein to develop an organ that functions similarly across many mammals. Biologists call this sort of thing 'convergent evolution'. There is certainly a bigger game afoot; the role played by pathogens such as viruses remain mysterious and elusive.

In Brief

In a series of lectures in Trinity College, the brilliant physicist Erwin Schrödinger laid down ideas that would irrevocably shift our views on biology. Inspired by Schrödinger, the discovery of the DNA spurs the blending of biology and computers – heralding the new era of systems biology. Prions not only cause mad cow disease but also aid the development of long-term memory. Equipped with the knowledge of the DNA, genetic engineering is used to efficiently produce medicine e.g., insights on how closely our genes are related to the functioning of our brains. The new CRISPR technique further catalyzes progress in genetic engineering. Using state-of-the-art techniques, genetic technology is employed to solve global issues from the energy crises to global food supply. With the aid of techniques like CRISPR, we take a peek into the inner-workings of the DNA – especially bits that are 'viral'. The human placenta is created by viral genes known as ERVs.

References

Ackerman, J., 2002. Food: How Safe. *National Geographic,* 201(5), pp.2-31.

Al-Khalili, J. and McFadden, J., (2015). *The quantum weirdness powering biology*, http://www.labnews.co.uk/

Alon, S., Garrett, S.C., Levanon, E.Y., Olson, S., Graveley, B.R., Rosenthal, J.J. and Eisenberg, E., 2015. The majority of transcripts in the squid nervous system are extensively recoded by A-to-I RNA editing. *Elife,* 4, p.e05198.

Atlantic, (2007). *Encyclopedia of Information Technology,* Atlantic Publishers & Distributors.

Appleby, B.S. and Lyketsos, C.G., 2011. Rapidly progressive dementias and the treatment of human prion diseases. *Expert opinion on pharmacotherapy,* 12(1), pp.1-12.

Azvolinsky, A., 2016. Arming Synthetic Bacteria Against Cancer, *The Scientist*:http://www.the-scientist.com/?articles.view/articleNo/46595/ title/Arming-Synthetic-Bacteria-Against-Cancer/

Badran, A.H., Guzov, V.M., Huai, Q., Kemp, M.M., Vishwanath, P., Kain, W., Nance, A.M., Evdokimov, A., Moshiri, F., Turner, K.H. and Wang, P., 2016. Continuous evolution of Bacillus thuringiensis toxins overcomes insect resistance. *Nature,* 533(7601), pp.58-63.

Bechtel, K. and Geschwind, M.D., 2013. Ethics in prion disease. *Progress in neurobiology,* 110, pp.29-44.a

Bogoch, I.I., Brady, O.J., Kraemer, M.U., German, M., Creatore, M.I., Kulkarni, M.A., Brownstein, J.S., Mekaru, S.R., Hay, S.I., Groot, E. and Watts, A., 2016. Anticipating the international spread of Zika virus from Brazil. *Lancet* (London, England), 387(10016), pp.335-336.

Boisvert, C.A., 2005. The pelvic fin and girdle of Panderichthys and the origin of tetrapod locomotion. *Nature,* 438(7071), pp.1145-1147.

Bonnet, J., Yin, P., Ortiz, M.E., Subsoontorn, P. and Endy, D., 2013. Amplifying genetic logic gates. *Science,* 340(6132), pp.599-603.

Briggs, H., (2015). *US approves genetically modified salmon for food, Science and Environment,* BBC news: http://www.bbc.com/news/ science-environment-34869556.

Callaway, E., 2016. CRISPR's hopeful monsters: gene-editing storms evo-devo labs, *Nature,* doi:10.1038/nature.2016.20449: http://www.nature.

com/news/crispr-s-hopeful-monsters-gene-editing-storms-evo-devo-labs-1.20449#ref-link-1.

Chandler, D.L., (2015). *Quantum physics meets genetic engineering,* MIT News: http://news.mit.edu/2015/quantum-physics-engineered%20 viruses-1014.

Chuong, E.B., Elde, N.C. and Feschotte, C., 2016. Regulatory evolution of innate immunity through co-option of endogenous retroviruses. *Science,* 351(6277), pp.1083-1087.

Clark, D.P. and Russell. L.D., 2000. *Molecular Biology Made Simple and Fun.* 2nd ed. Vienna, IL: Cache River Press.

Coghlan, A., (2016). *Bio coding language makes it easier to hack living cells,* Daily News, New Scientist.

Cox, M.M. and Battista, J.R., 2005. Deinococcus radiodurans—the consummate survivor. *Nature Reviews Microbiology,* 3(11), pp.882-892.

DeWeerdt, S.E., (2002). The World's Toughest Bacterium: Deinococcus radiodurans may be a tool for cleaning up toxic waste and more, Genome News Network: http://www.genomenewsnetwork.org/articles/07_02/deinococcus.shtml.

Din, M.O., Danino, T., Prindle, A., Skalak, M., Selimkhanov, J., Allen, K., Julio, E., Atolia, E., Tsimring, L.S., Bhatia, S.N. and Hasty, J., 2016. Synchronized cycles of bacterial lysis for in vivo delivery. *Nature,* 536(7614), pp.81-85.

Dupressoir, A., Marceau, G., Vernochet, C., Bénit, L., Kanellopoulos, C., Sapin, V. and Heidmann, T., 2005. Syncytin-A and syncytin-B, two fusogenic placenta-specific murine envelope genes of retroviral origin conserved in Muridae. *Proceedings of the National Academy of Sciences of the United States of America*, 102(3), pp.725-730.

Elitzur, T., Yakir, E., Quansah, L., Zhangjun, F., Vrebalov, J.T., Khayat, E., Giovannoni, J.J. and Friedman, H., 2016. Banana MaMADS transcription factors are necessary for fruit ripening and molecular tools to promote shelf-life and food security. *Plant physiology,* pp.pp-01866.

Fang, G., Rocha, E. and Danchin, A., 2005. How essential are nonessential genes?. *Molecular biology and evolution,* 22(11), pp.2147-2156.

Feldman, D., 1979. Virus particles in the basal plate of rhesus monkey and baboon placenta. *Cancer research,* 39(5), pp.1772-1783.

Gassmann, A.J., Petzold-Maxwell, J.L., Clifton, E.H., Dunbar, M.W., Hoffmann, A.M., Ingber, D.A. and Keweshan, R.S., 2014. Field-evolved resistance by western corn rootworm to multiple Bacillus thuringiensis toxins in transgenic maize. *Proceedings of the National Academy of Sciences*, 111(14), pp.5141-5146.

Gill, V., (2010). 'Artificial life' breakthrough announced by scientists', *Science & Environment*, BBC News.

Hammond, A., Galizi, R., Kyrou, K., Simoni, A., Siniscalchi, C., Katsanos, D., Gribble, M., Baker, D., Marois, E., Russell, S. and Burt, A., 2016. A CRISPR-Cas9 gene drive system targeting female reproduction in the malaria mosquito vector Anopheles gambiae. *Nature biotechnology,* 34(1), pp.78-83.

Heaven, D., (2013). 'DNA transistors pave way for living computers', *Daily News*, New Scientist.

Heidmann, O., Vernochet, C., Dupressoir, A. and Heidmann, T., 2009. Identification of an endogenous retroviral envelope gene with fusogenic activity and placenta-specific expression in the rabbit: a new. *Retrovirology,* 6(1), p.107.

Henzy, J., (2014). *Retroviruses, the Placenta, and the Genomic Junk Drawer*: http://schaechter.asmblog.org/schaechter/2014/06/retroviruses-the-placenta-and-the-genomic-junk-drawer.html.

Hutchison, C.A., Chuang, R.Y., Noskov, V.N., Assad-Garcia, N., Deerinck, T.J., Ellisman, M.H., Gill, J., Kannan, K., Karas, B.J., Ma, L. and Pelletier, J.F., 2016. Design and synthesis of a minimal bacterial genome. *Science,* 351(6280), p.aad6253

Jones, J.D., Witek, K., Verweij, W., Jupe, F., Cooke, D., Dorling, S., Tomlinson, L., Smoker, M., Perkins, S. and Foster, S., 2014. Elevating crop disease resistance with cloned genes. *Philosophical Transactions of the Royal Society of London B: Biological Sciences,* 369(1639), p.20130087.

Kingery, K., 2016. *Directly reprogramming a cell's identity with gene editing: Engineers use CRISPR to generate neuronal cells from connective tissue*, Duke University.

Li, W., Lee, M.H., Henderson, L., Tyagi, R., Bachani, M., Steiner, J., Campanac, E., Hoffman, D.A., von Geldern, G., Johnson, K. and Maric, D., 2015. Human endogenous retrovirus-K contributes to motor neuron disease. *Science translational medicine,* 7(307), pp.307ra153-307ra153.

Luntz, S., (2015). *Squid Can Recode Their Own Genetics,* http://www.iflscience.com/plants-and-animals/squid-recode-their-own-genetics/

McKay, B., 2016. Mosquitoes Are Deadly, So Why Not Kill Them All?, *The Wall Street Journal*:http://www.wsj.com/articles/mosquitoes-are-deadly-so-why-not-kill-them-all 1472827158.

McNamee, E. and Wilson, J., 2013. *A Nobel Prize with help from sea slugs,* http://edition.cnn.com/2013/05/14/health/lifeswork-eric-kandel-memory/

Nakamura, T., Gehrke, A.R., Lemberg J., Szymaszek J., Shubin N.H., Digits and fin rays share common developmental histories, *Nature,* 2016, Vol 537(7619), pp 225-228.

Nielsen, A.A., Der, B.S., Shin, J., Vaidyanathan, P., Paralanov, V., Strychalski, E.A., Ross, D., Densmore, D. and Voigt, C.A., 2016. Genetic circuit design automation. *Science,* 352(6281), p.aac7341.

Okabe, M., Ikawa, M., Kominami, K., Nakanishi, T. and Nishimune, Y., 1997. 'Green mice'as a source of ubiquitous green cells. *FEBS letters,* 407(3), pp.313-319.

Park, H., Heldman, N., Rebentrost, P., Abbondanza, L., Iagatti, A., Alessi, A., Patrizi, B., Salvalaggio, M., Bussotti, L., Mohseni, M. and Caruso, F., 2015. *Enhanced energy transport in genetically engineered excitonic networks. Nature materials*.

Pocchiari, M., Puopolo, M., Croes, E.A., Budka, H., Gelpi, E., Collins, S., Lewis, V., Sutcliffe, T., Guilivi, A., Delasnerie-Laupretre, N. and Brandel, J.P., 2004. Predictors of survival in sporadic Creutzfeldt–Jakob disease and other human transmissible spongiform encephalopathies. *Brain,* 127(10), pp.2348-2359.

Sample, I., (2010). Craig Venter creates synthetic life form, *The Guardian*. https://www.theguardian.com/science/2010/may/20/craig-venter-synthetic-life-form.

Tompa, R., 2016. CRISPR and beyond: The ins and outs of gene editing and its potential for cures, *Hutch News*, https://www.fredhutch.org/en/news/center-news/2016/08/crispr-and-other-gene-editing-show-cure-potential.html.

Unckless, R.L., Clark, A.G. and Messer, P.W., 2016. *Evolution of resistance against CRISPR/Cas9 gene drive*, http://biorxiv.org/content/early/2016/06/11/058438. Abstract.

Von Neumann, J. (1948). *The Computer and the Brain.* New Haven and London: Yale University Press.

Wildschutte, J.H., Williams, Z.H., Montesion, M., Subramanian, R.P., Kidd, J.M. and Coffin, J.M., 2016. *Discovery of unfixed endogenous retrovirus insertions in diverse human populations.* Proceedings of the National Academy of Sciences, p.201602336.

Web Sources

An Overview of the Human Genome Project, 2016. https://www.genome.gov/12011238/an-overview-of-the-human-genome-project/

Columbia University Medical Center (CUMC), 2015. "Long-term memories are maintained by prion-like proteins." *Science Daily*, 2 July 2015. <www.sciencedaily.com/releases/ 2015/07/150702132804.htm>.

Gardasil (HPV), (2006), http://www.fda.gov/BiologicsBloodVaccines/Vaccines/QuestionsaboutVaccines/ucm096052.htm.

Insulin, 2008, http://www.madehow.com/Volume-7/Insulin.html.

WHO, Micronutrient deficiencies, Vitamin A deficiency, (2016): http://www.who.int/nutrition/topics/vad/en/

Chapter 4

PHYSICS OF LIFE

"Somewhere, something incredible is waiting to be known."
— Carl Sagan

LITTLE DIFFERENCES MATTER

In the early fifties, while running climate simulations on a computer, a young weatherman at MIT discovered a strange phenomenon which would significantly change the directions of scientific progress across a wide range of disciplines. This young weatherman was Edward Lorenz and the phenomenon in question is known today as chaos theory. After obtaining a Master's degree in mathematics from Harvard, Lorenz joined the U.S. Army Air Corps. In the army Lorenz used his mathematical skills for weather forecasting. Interested in the weather, Lorenz enrolled into MIT obtaining a second Master's degree and consequently a doctorate in meteorology. Working as a professor at MIT, Lorenz inserted some parameters into his computer simulator and left it to run while he left his office to get some coffee. The parameters defined in the simulator were weather variables such as temperature and wind speed. Due to his strong math training, Lorenz developed and wrote various formulations and computer code for weather prediction. During this particular run Lorenz rounded-off a parameter from 0.506127 to 0.506 (three decimals). Back from his coffee break he was astonished to see that the results churned out from his simulator significantly differed due to the rounding-off. He observed a whole new predicted weather pattern over a time scale of two months. This phenomenon is known as the 'butterfly effect' – where small changes could cause large impacts.

The term 'butterfly effect' came from the common example often used to describe chaos theory; a tiny flap of a butterfly wing in one corner of the globe could cause a tornado on the other side. As with the butterfly, chaos theory extends its grasp broadly throughout many scientific areas including geology and biology. In the eighties, physicists came to realize that chaos theory went up against our classical concepts of nature. The deterministic and predictable mechanics promoted since the time of Newton and Galileo appeared to be under question. This is of course not the first time classical physics has been challenged thanks to quantum physics and the theory of relativity. As put elegantly by the author Michael Crichton, in his novel 'Jurassic Park' – "They believed that prediction was just a function of keeping track of things. If you knew enough, you could predict anything. That's been the cherished scientific belief since Newton. And? Chaos theory throws it right out the window." In classical physics we learn that if we know the initial conditions of any system, then predicting its behavior in the future shouldn't be much of a problem. With chaos theory we have come to realize that classical situations work for systems with short periods. When these systems are studied for lengthy time intervals, chaotic behavior starts to manifest – where slight variations in the initial settings could propagate into huge effects in the future, making things very unpredictable. The weather systems researched by Lorenz are greatly sensitive to initial conditions. This makes predictions too far ahead into the future difficult.

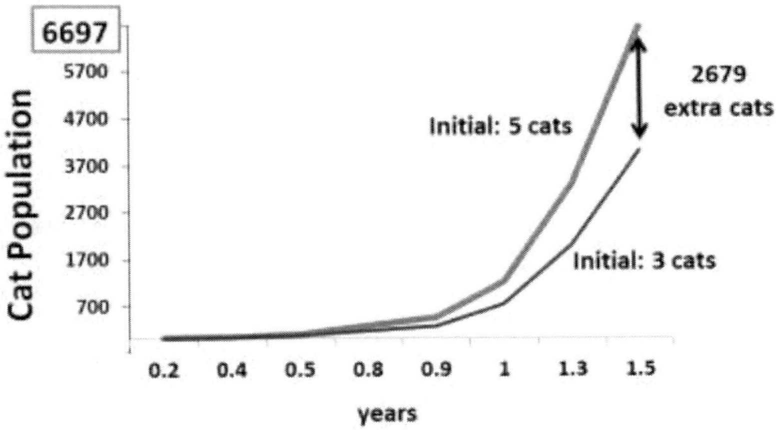

Figure 4.1. Chaotic Growth of a Cat Population.

The findings of Lorenz show how nonlinearity (resulting in unpredictability) starts to creep in when a system is let to evolve for long periods (Dizikes, 2011). Take population growth for instance; imagine an initial population of 3 cats growing at pace where they double every 7 weeks (10% growth rate). In one and a half years the cat population becomes about 4018. Instead if we start with an initial population of 5, then we get the final population of a staggering 6697 cats. Just adding 2 cats to the initial population caused the population to vary by 66.6% within one and a half years (Figure 4.1). This is because populations grow nonlinearly and when stretched over long periods, chaos sets in; making the system highly sensitive to initial conditions.

Since its discovery, chaotic patterns have been emerging steadily across the life sciences. Chaos-based ideas have been seen to daringly challenge core concepts in Darwin's theory of evolution. The two fundamental ideas presented by Darwin are:

1. Common Ancestor Concept.
2. Selection and Adaptation.

The first concept states that life evolves from common ancestors – meaning that if we trace down the ancestral lineage of various organisms, we will eventually reach common ancestors. The second idea known as 'adaptionism' which explains changes in species as it adapts to changes in the environment. For instance, Great White sharks adapted to locating hidden prey by evolving electroreceptors to detect minute electric fields emitted by the prey's muscles (Berquist et al., 2015). Adaptionism is very effective for explaining microevolution, the evolution within a single species or in small groups. However, when applied to large groups of various species, over lengthy time periods (macroevolution), adaptionism starts getting wobbly (Bennett, 2010). A scenario Darwin did not foresee in his day; thinking that evolutionary change and adaptation remain continuous and gradual over time regardless the scale.

Thanks to steady scientific progress in the past years, scientists managed to amass a good amount of data of fossil record over the past two million years. Using this data they could further test Darwin's adaptionism idea by asking the question: how did life react to climate change in the last, say twenty thousand years? To answer this scientist started examining fossilized tree pollen from lake and peat sediments. Using carbon-dating, they identified the genus and even species of tree fossils. They found that most of the tree fossils show no significant reaction to climate change (Davis, 1983). Although they do respond individually, they don't show any critical changes across groups of species – proving that they do not evolve in groups as taught by Darwin's theory. When scientists took a look at fossilized insects, they surprisingly found similar 'no effect' trends even when they were facing many ice ages (Coope, 2004). These findings stress the weakness of the link between large-scale evolution and changes in the environment.

So how does chaos influence evolution of large groups of species (or macroevolution)? We've seen how slight changes in initial numbers could produce large impacts to population sizes in the long run. This goes the same for the genome. In many organisms a single mutation in the genome might not cause significant changes to the individual organism. As always, chaos begins its work after many generations of reproduction. The initial mutation then starts accumulating more mutations in the subsequent generations; gradually becoming nonlinear. Then large changes occur in the environment forcing the organism to react. Manifesting from the genome, this reaction influences the organism's characteristics (or phenotype). Therefore, this interplay between the genome and the phenotype is chaotic and unpredictable – where the organism responds in a chaotic manner by altering its phenotype based on its genome. This chaos viewpoint throws itself head-on towards Darwin's theory of gradual adaptation and evolution. Although gradual adaptation may work on individual species, the theory suffers when faced with large-scale evolution happening over lengthy periods. The macroevolution idea thus falls into the realm of chaos theory.

Bottom line is; when there is a glacial event, catastrophe or rapid climate change, organisms' don't react and evolve in a gradual and continuous manner. Their evolution becomes chaotic and unpredictable – depending on the accumulation of mutations (nonlinear) in their genome and their sensitivity to the environment. If the organisms are very sensitive to the environment and mutations get triggered accordingly, then significant evolution occurs. Otherwise nothing or minimal evolution occur accumulating minor mutations – lying in wait for the conditions to be just right.

When we say something exhibits chaos, it means that it's sensitive to initial conditions and may 'seem' unpredictable. Although chaotic systems may look unpredictable, they are actually predictable to a certain degree – their behavior is not completely random. Additionally the source of chaos (or equilibriums) could be identified. Biologists and researchers are bumping into chaotic systems in many of their studies. Interestingly, the human heart even when beating healthily is slightly chaotic – though not chaotic enough to spin out of control. The heart operates on the fringes – where order meets chaos (Ives, 2004). Biologists say that a slightly chaotic heart is more capable for coping with sudden and large variations in blood flow. Imagine having a gradual or linearly pumping heart – if there is a sudden change in the body (e.g., you get thrown off the couch while reading this book), your heart would need to beat faster to supply more blood to different parts of your body for a speedy recovery from your fall. A linear heart would not be able to compensate in good time for such a rapid demand. Of course a completely chaotic heart would simply beat erratically inducing a full-blown heart attack. Hence, a slightly chaotic heart is what does the trick giving the heart the right amount of unsteadiness so that it can respond speedily. This way you can pick yourself up and recover from the fall in a short time. Recently researchers have found that the heart's rhythm before a cardiac arrest exhibits increased chaotic behavior (Dai and Schaeffer, 2010). As it gets more chaotic, the heart becomes more erratic leading to a complete cardiac arrest. The model may help researchers to design a detector to detect the onset of a heart attack in advance. This heads-up would give patients enough time to call for medical help right before an attack.

Besides the heart, even the brain operates by using a little help from chaos. In 2014, Canadian researchers developed methods to predict the onset of seizures. Before looking at seizures, the team focused on using their computer program for predicting other stuff like the financial market and weather. The idea was to use the level of chaos as a marker to predict future behavior of the systems they were studying. The computer program could be thought of as a radio with a knob for tuning the level of chaos (instead of the correct station frequency). By tuning the computer program for the correct level of chaos, they manage to make accurate predictions of the system they were studying. Using right level of chaos, the team predicted if the brain was operating normally or tipping over to an epileptic seizure (Golestani and Gras, 2014). As with seizure prediction, scientists are also looking into chaos theory for detecting conditions like Parkinson's disease and migraines (Yulmetyev et al., 2006; Bosl et al., 2011; Kernick, 2005). Although widespread, their true sources still elude doctors to this day. Chaos–related diseases are now called 'dynamical diseases'.

SOUP FOR STARTERS

Alexander Oparin, a Russian biochemist was well known for developing one of the earliest theories of how life originated on earth. Oparin's theory was what we call today the 'primordial soup' theory. According to this theory, life evolved from carbon-based chemical molecules. Working in Russia, Oparin was honored with many awards for his scientific works, so much so that he was sometimes referred to as the "Darwin of the Twentieth Century." In Oparin's theory life evolved from carbon-based compounds in a kind of primordial soup. He believed that life could spring via evolution from lifeless (inanimate) matter. In Oparin's idea, the early Earth had a chemically reactive atmosphere consisting of methane, hydrogen sulfide, ammonia, carbon dioxide/monoxide, phosphate and water vapor along with a little oxygen. He considered these basic chemicals to be the building blocks of life. During multiple reactions, these molecules continually react among themselves achieving greater complexity. As the molecules grow in complexity their characteristics become more multifaceted. Once enough complexity is achieved, life manifests itself and the process of evolution, speciation and so on continues. Oparin took Darwin's theory and projected it backwards in time. His theory that life came into being by chance was initially met with strong resistance and skepticism. Nevertheless, Professor Harold Urey and his student Stanley Miller at the University of Chicago took a keen interest in Oparin's idea. So keen they wanted to test it out.

According to Oparin, the chemical compounds would accumulate and react to form long molecular chains similar to nucleic acids and proteins (found in almost all life forms on Earth). These nucleic acids carry information essential to the operations of any organisms and are stored in strands of DNA (or RNA). In 1953, Urey and Miller built an experimental setup to simulate the atmospheric conditions of early earth containing Oparin's primordial soup. In addition to the basic chemicals, the setup was equipped to provide the primordial soup with sudden bursts of electricity for simulating lightning storms (in the spirit of Dr Frankenstein's apparatus). The Miller-Urey experiments managed to produce 5 amino acids. When this experiment was repeated using gases like those occurring in volcanic eruptions, researchers found a total of 22 amino acids (Johnson et al., 2008). Unfortunately, along with the results Miller and Urey could not explain how the basic compounds formed amino acid chains. Nevertheless the results had a colossal impact on our understanding of the origins of life on Earth. Taking into account some recent developments, there is still no other origin of life theory standing as strong as the Miller-Urey experiments.

Spurred by Miller and Urey, the following decades saw scientists continue their search for the origins of life. In 2010, a research group lead by Nick Lane at University College London proposed an alternative theory stating that life originated from hydrothermal vents from the earth's oceanic floor – "We present the alternative that life arose from gases (hydrogen, carbon dioxide, nitrogen, and hydrogen sulfide) and that the energy for first life came from harnessing geochemical gradients created by mother Earth at a special kind of deep-sea hydrothermal vent – one that is riddled with tiny interconnected compartments or pores", claimed Nick Lane (Wiley-Blackwell, 2010). Deep sea vents are alkalic and could produce chemicals used by organisms for energy (known as chemiosmosis). Using this energy the team argue that organisms came into being synthesizing their own chemicals and evolving as they go on.

A more recent physical theory of the origin of life was developed by Jeremy England, a biophysicist working at MIT. To understand England's version, let's revisit the concept of 'entropy' presented in the initial part of Chapter 3. Entropy was among the key physics concepts used by Schrödinger to distinguish life apart from other inanimate forms. Entropy represents the level of 'disorder' in systems. In the Trinity Lectures, Schrödinger stated that unlike inanimate systems (which always progress to higher levels of entropy), living things could reverse the process; moving towards lower degrees of entropy. Instead of decaying away, life forms could build and maintain order in their structure. This way the living progress achieving higher order with lower entropy states. England derived a mathematical formula depicting this strange entropy-reversing capability of life forms (England, 2013). Using this formula, England shows that if a cluster of inanimate matter were left exposed to an energy source (e.g., heat), this cluster would start organizing itself striving to achieve higher order. This way it lowers its entropy by dissipating more heat as it self-organizes, gradually becoming more complex. At some stage during this growth, it becomes complex enough to be considered alive. In England's origin of life theory, any clump of matter exposed to certain surrounding conditions could spring into life – 'You start with a random clump of atoms, and if you shine light on it for long enough, it should not be so surprising that you get a plant' (Wolchover, 2014). England's idea seems to strengthen results presented by the Miller-Urey experiments.

Concepts like entropy come from the branch of physics called thermodynamics or the physics of heat. Among the pillars of thermodynamics is something called the Second Law – which states that all matter must decay towards a higher entropy (greater disorderness). Although Schrödinger pointed out that living forms have the capability to reverse entropy, they cannot keep doing this forever. The living cannot maintain their material organization forever avoiding decay and becoming eternal. At some point they eventually lose their capability of reversing entropy – they grow old and perish. The overall entropy does get higher with time as the organism grows old. And all things living or otherwise abide the Second Law of thermodynamics. The key concept England proposes is the ability to dissipate energy effectively; which reverses entropy preserving the material organization of the organism. This preservation prolongs its life. If we think of this theory as a basis, then Darwin's evolutionary picture would seem like the most efficient way for organisms to dissipate energy with respect to certain surrounding conditions – this provides a physical foundation underlying Darwin's evolution of species.

Keeping It Organized

To dissipate energy effectively and fend off entropy, organisms would have to keep restructuring, repairing and keeping fit. In other words, they would have to 'self-organize'. Self-organization has been seen to occur in diverse scenarios and varying forms in all circles of life. Self-organization is found frequently in insect swarms – among the mostly studied types are the ones which happen in ant colonies. Being highly social, ants tend to form strong bonds among members of their colony. In ant colonies, 'the needs of the many outweigh the need of the few' (as famously put by Spock in the 1982 film, Star Trek). This description is most fitting since individual ants would gladly throw themselves in the way of danger for the sake of the colony. When hit by floods, fire ants assemble into floating rafts by linking their mouths and legs together (Howard, 2015). The raft consists of up to 100, 000 individual ants. Once the ants weave their raft (in less than two minutes by the way), they immediately position the queen and their young larvae in the center – securing them high and away from the water. Fire ants could survive for weeks on these rafts – where the ants forming the raft face the dangers of being eaten by fish or other predators. For these ants, the sacrifice is worth making and the risk is worth taking.

This raft-building business is self-organization at its best. Each individual ant per se is not intelligent enough to design and carry-out such feats. These ants operate without a leader or preplanned design (single ants are not very intelligent). On the other hand, if each individual ant equipped only with simple rules interact in a large group, they cumulatively act similar to a single highly intelligent organism. Joining intelligence, they self-organize to come up with complex responses when facing challenges posed by the environment. Besides rafts, ant colonies have also been observed to build bridges.

Army ants in the amazon build bridges across their destinations to expedite their travels. These ants travel out of their nests to ransack other insect colonies for food. The food is then brought back to feed their young. Similar to forming rafts, bridge building results from simple rules practiced by single ants. In light of the research findings of Christopher Reid and his team from the University of Sydney, we now have some idea what these rules are and how they are used by army ants for construction (Reid et al., 2015). Reid's team ran experiments on army ants using pheromones. Pheromones are hormones which are usually deposited by single ants to communicate with others. High levels of deposited pheromones show that many ants have travelled the path – deeming the trail preferable for future use. Low levels of pheromone signify less preferable or even perilous trails. Hence, pheromones are used by ants as a form of 'chemical communication' (see Figure 4.2).

Reid's team ran these experiments in Panama; where they directed ants to follow zigzag paths using pheromones. They observed that first one ant would volunteer to use its body as a bridge to connect a small gap between the zigzag trails. This slight improvement is followed by more ants connecting to the bridging ant to improve the angle of the path. They keep improving trying to achieve a shorter path. The ants which are part of the bridge with greater distance will disconnect and continue to march. This way the bridge keeps moving towards an optimal distance, shortening the travel path. Upon reaching close to the shortest possible travel length, the ants stop improving the bridge. They apparently know the point when they have achieved maximum food gathering efficiency. According to Reid's group, individual ants decide when to disconnect from the bridge based on how frequently they get touched by other ants – which translates to how often the ants are using that path. If many ants walk over them, then they stay connected holding the bridge. Otherwise they let go and continue their trail. Using only this simple rule, army ants build and optimize their bridges to expedite their travel. Army ants could even build temporary nests. When needed, they construct structures called 'bivouacs' using their own bodies. These bivouacs protect their young as well as the queen ant.

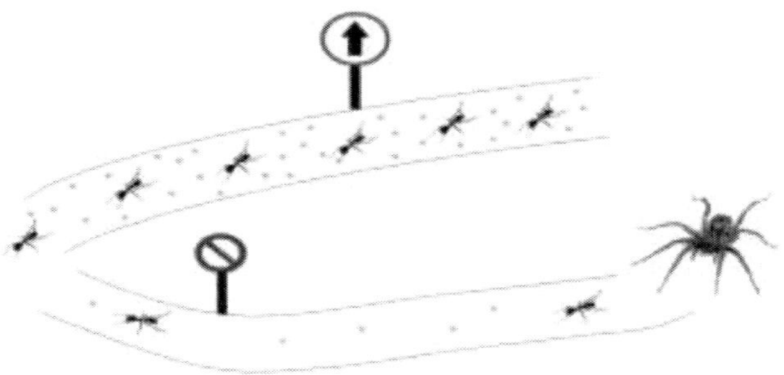

Figure 4.2. Pheromone Communication among Ants.

Just to give us a feel of the kind of intelligence we are talking about here, let's take a look at the number of neurons in the brains of some animals. The number of neurons don't directly depict intelligence, it's the way they are connected (neural architecture) that relates to intelligence. For example, large animals like elephants have much higher number of neurons as compared to chimpanzees, but this doesn't make them more intelligent by comparison. Excluding large animals, we can still draw a crude picture of intelligence by comparing the number of neurons. So let's give it go: we have humans with about 86 billion neurons, Guinea pigs score at 240 million and frogs have 16 million neurons. Cockroaches pack 1 million neurons while ants have a meager 250, 000 neurons. With 250, 000 neurons, ants build structures like bridges and rafts – they even create complex massive underground ant cities (Monaenkova et al., 2015).

INSECT ENGINEERS

Using self-organization, insect colonies achieve feats of complex engineering. Discovering complex underground nests, scientists couldn't help but wonder how these ants maintain suitable underground temperature and ventilation necessary for nurturing their young. Recently, with the help of lab experiments, scientists found that ant colonies build turret-like structures directly above their subterranean homes. Riddled with pores, these structures allow air to flow into the chambers of their nests (Cosarinsky et al., 2012). Air ventilation systems ensure the necessary conditions for ants to farm their fungus – besides foraging for food ants do of bit of agriculture on the side to produce fungus for feeding their young. Termites too engineer intricate structures that rival their ant counterparts. Like ants, termites have high metabolism rates – they consume high amounts of energy in short periods of time. To maintain this metabolism they require lot of oxygen. This means that similar to ants they need pretty good ventilation systems transporting adequate oxygen to the chambers in their subterranean nests. Besides, waste carbon dioxide should be disposed from their nests as quickly as the intake of oxygen, or else the colony will suffocate.

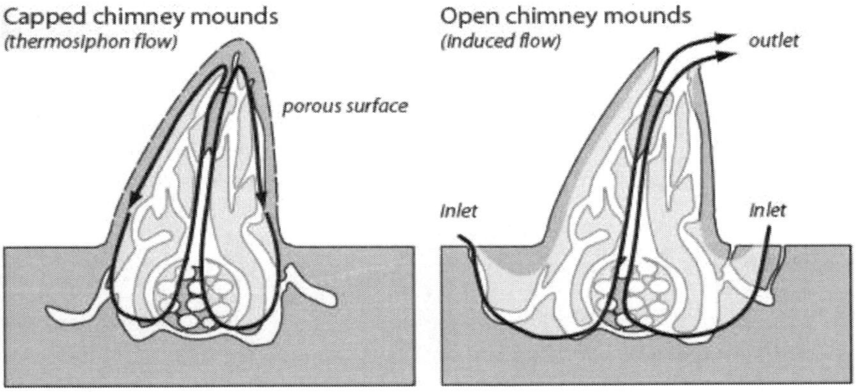

Figure 4.3. Architecture of Termite Mound ((Turner and Soar, 2008).

The nest seems to breathe and consume oxygen similar to the workings of our own lungs (Turner and Soar, 2008). Previously, it was thought this breathing mechanism was driven by the winds from outside the mound. This is because unlike our lungs, mounds don't have muscles to drive it to inhale and exhale. Researchers have found that the mound's ventilation is driven by outside temperatures (King et al., 2015). The mounds are constructed with a large tube in the center acting like a chimney (Figure 4.3). This chimney then splits into smaller tubes connecting to the chambers in the mound. During daytime, air is warmed causing warm air to rise from the smaller tubes sinking into the chimney keeping the chambers nice and cool. At night, the process is reversed where warm air from the chimney moves to the chambers keeping it warm. This intricate underground mound architecture not only regulates the mound's internal temperature, but ensures that the required amount of oxygen and carbon dioxide are delivered and expelled according to the needs of the resident termites.

The biologist, Guy Theraulaz from the University of Toulouse tried to verify the hypothesis that complex behavior observed in insects emerge from self-organization (Perna et al., 2012). Theraulaz and his team set out to do this by developing a computer model which operates on the basis of three simple rules:

1. Each ant picks up sand grains at a constant rate.
2. An ant prefers to drop grains near other grains.
3. Grains handled by more ants are favored to be picked up as opposed to grains handled by fewer ants (pheromone marking method).

When they ran simulations using the computer model, the team found that they could mimic nest-building behavior of real ants. Impressed with the results, the researchers attempted to introduce environmental factors into their computer model to simulate conditions such as heat and humidity. Under the new conditions, the virtual ants emulated their real counterparts building similar nest structures. Such computer models certainly make a strong case for using concepts like self-organization to explain complex behavior of insects like ants; as explained by Theraulaz, 'The real novelty here is our newly acquired ability to observe in detail the formation and the transformations of these structures. We finally have access to precise data on how living things get together to form complex yet fully functional and reactive structures' (Singer, 2014a).

Valuable Feedback

Whether in ants or termites, self-organization always leads to increased complexity. Starting from simple rules each member interacts making the system more complicated. This makes them capable of generating complex responses like engineering mound ventilation systems and building bridges. In order for self-organization to successfully occur, the individual members will not only need to receive signals from other members, but they would also need to send signals – making the communication two-way. Therefore, each individual member provides an appropriate feedback depending on the kind of signal they receive. Take the bridge-building ants for example – if the ant forming the bridge does not feel other ants stepping on its body, it immediately disconnects from the bridge and continues foraging. On the other hand, feeling other ants walking over it makes it stay connected to the bridge.

One type of feedback commonly experienced by most people is the screeching sound of audio feedback. This happens when a microphone and speaker get caught in loop. The sound from the microphone is amplified by the speaker and transmitted. The transmitted sound is picked by the microphone again and is amplified by the speaker. This loop repeats creating a loud irritating screech. This happens because the feedback signal is passed from the speaker to the microphone and back to speaker again forming a loop.

Although not particularly useful in audio systems, feedback responses play a critical role in many of nature's self-organizing systems. With many individuals responding to each other, high complexity and self-organization is achieved. Working at Stanford University, the biologist Deborah Gordon ran a series of interesting experiments on ants proving the importance of their feedback to the survival of their colonies (Gordon et al., 2013; Prabhakar et al., 2012). Gordon's team placed cameras recording ants moving from an opening in their nest to a food source nearby. These ants move back and forth carrying food supply to their nest. The team then removed some ants carrying food on their way back to the nest. When they did this, the ants based at the nest stopped heading out to the food source. The scientists found that the ants heading back from the food source provided the feedback required for other ants to go out to the food source – like a relay race. What's happening here is called positive feedback. Positive feedback encourages the ants to go out and collect food. Too much of positive feedback, however, endangers the ants – as they could all end up heading out at one go risking attacks by predators or wasting precious travelling when food supply is already exhausted. To avoid this from happening, many self-organizing systems have negative feedback. Negative feedback mellows down the ants' response based on the availability of food supply. As the food supply diminishes, lesser ants gradually head out. This way the ants don't waste resource and risk their lives. By picking-off ants traveling back to their nest from the food supply, Gordon's team artificially simulated negative feedback causing the ants back in the nest to stop heading out from their nests – signaling danger up ahead. The balance between positive and negative feedback is the yin and yang of self-organization. The balance is designed to ensure the on-going survival of members in the group.

Besides ant colonies, feedback mechanisms are also found in the brain and nervous system. Each neuron in the brain sends a signal to the neuron next to it – influencing the potential for this second neuron to signal the next one. Some neurons (excitors) enhance the signal to the next neuron while others function as inhibitors, reducing the signal. A combination of two types of neurons enables the neighboring neuron to decide if it's going to signal or not. And if it signals, it needs to decide if it's going to amplify or inhibit the signal passing through to the next neuron. However, once the neighboring neuron has acted (excited/inhibited the signal), it provides a feedback signal to the previous neuron which stimulated it in the first place – forming a neural feedback loop. Therefore, the initial neuron now knows how the neighboring neuron has acted. With this information it then acts accordingly. This way you have a group of neuron amplifying (positive feedback) or inhibiting (negative feedback) signals. Scientists at Yale University found that they could artificially induce positive (or negative) feedback in animal brains during sleep. They do this by applying weak electric fields (Fröhlich et al., 2010; Arnold, 2015). This way they could amplify and inhibit signals already existing in the animal brain. The experiments show that neurons (like ant colonies) use feedback loops to function. In psychology, the idea of loops and cycles are not alien. Psychologists often use terms like manic cycles and toxic loops to describe mental disorders. Perhaps this is not so surprising since at a fundamental level neurons work using feedback loops as a basis.

The Attractor Factor

Feedback loops are crucial components that add complexity to self-organized systems. Some complex systems behave in a chaotic manner manifesting chaotic characteristics such as being sensitive to initial conditions. Another important feature of complex systems is the existence of 'attractors'. Attractors are certain states a system always gravitates towards. Although starting at different points, systems often get 'attracted' towards these states. For instance, the cat population in Figure 4.1 has two attractors. Then for the initial population of 5 to 20 cats, after a fixed period of time the population always reaches around say 5000 cats. However, the moment the initial population tips over 20 cats, suddenly (after the same period) the cat population jumps to about 10000. In such cases, it is said that the cat population system has 2 attractors for 2 ranges of initial conditions.

Another simple case of a two-attractor system is a surface with two holes; as shown in Figure 4.4. Imagine letting a ball roll on such a surface with different starting points. If the ball starts rolling at any point before the critical line, it will eventually drop into Hole A. But if the ball starts rolling anywhere after the critical line (see dashed line in Figure 4.4), then it will fall into hole B. For a range of initial conditions (starting points), the ball falls into one hole – and if it crosses that range, it falls into the next one. The holes could be thought of as attractors where a system can end-up depending on its starting range.

Existence of multiple attractors cause systems to have critical lines or 'tipping points'. Attractors also occur in social settings as presented by psychologist, Malcolm Gladwell in his bestseller book, 'Tipping Point' (Gladwell, 2000). In the book, attractors are used to explain the Baltimore syphilis epidemic. The nineties saw a sudden and sharp increase of children born with syphilis. It was as though the disease has crossed a critical line 'dropping' into another new attractor. The Centre for Disease Control (CDC) claimed that the cause for this sudden rise in syphilis was crack cocaine. Increased use of the drug caused people to become more susceptible to risky sexual behavior. This resulted in more children born with the disease. Some experts suggested that the epidemic was caused by the collapse of the healthcare system due to radical budget cuts by hospitals and the scaling down of STD clinics. The epidemiologist, John Potterat had a completely different explanation – he pointed to housing demolition projects. At first glance, the idea seems far-fetched, but not if seen using an 'attractor' perspective. According to Potterat, Baltimore city planned to demolish many of their old public housing (especially high-rises). Housing hundreds of families, these high–rises were crime-ridden and infested with various disease. While high-rises were being demolished, many people were leaving their declining terrace houses in the Eastern and Western parts of the city. Potterat was saying that for many years the disease was concentrated in these parts of the city. The sudden demolition of a few housing projects caused the disease to spread more rapidly throughout the city. There has been cases of syphilis in Baltimore before the mid-nineties. Increased abuse of crack cocaine and poor healthcare may have contributed to the widespread of the disease. During those times, syphilis infections were still within the critical line of one attractor. The demolition projects quickly tipped the system over into the domain of another attractor. The sudden sharp increase in syphilis cases is the system (epidemic) crossing this critical line, being subjected to the new attractor. Any complex system, be it a population of cats or disease epidemics, multiple factors could push it closer to the critical line. Then with a just little nudge, the system could easily tip over the critical line, falling into the domain of another attractor.

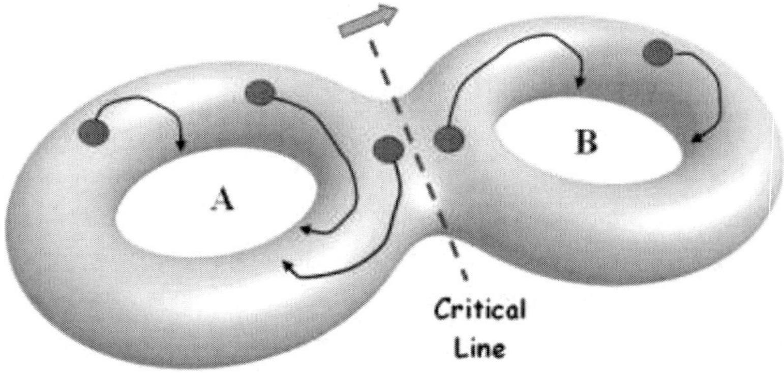

Figure 4.4. Two-holed Attractor.

CRITICAL LINES IN THE ECOSYSTEM

Inhabited by various species of plants and fauna, ecosystems are among the most complex natural networks existing on our planet. Such complexity inevitably gives rise to the manifestation of attractors. Some years back, Stephen Capenter, a biologist at the University of Wisconsin, Madison and his team conducted research revealing the sensitivity of ecosystems. They conducted what is widely known in ecological circles as the Peter Lake Experiment (Carpenter et al., 2010; Carpenter et al., 2011; Wolchover, 2015). Located in Wisconsin, Peter Lake's food chain consists of four main species: bass, minnows, fleas and algae. The top of the food chain is bass followed by flea-eating minnows. Sitting at the bottom of the food chain are algae which are food for flea. Carpenter's team studied the sensitivity of the lake's ecosystem relative to sudden changes in populations.

In these experiments they gradually increased the number of bass in the lake. They studied the effects by determining the population of other remaining members of the food chain. As expected, the minnows count started dropping due to increased predation by bass. This resulted in an increase in flea numbers causing an algae deficit. But they also did another analysis. They monitored how long it would take for the populations in the ecosystem to recover to its initial state – as they were prior to the addition of bass by the team. In each experiment, the team increased the number of bass and recorded the recovery time. They noticed that the more bass they introduced, the longer it took the lake to return to its initial state (longer recovery time). Carpenter's team calls this 'critical slowing down.' Every time a disturbance shifts the lake, internal forces in the ecosystem attract it back to its initial state – recovering it. However, as disturbances become stronger (more bass is introduced), the attracting forces weaken. This causes it to take a longer time to recover.

To see how long this recovery process could go on, Carpenter's team kept adding more bass into Peter Lake. They observed that when the bass populations increased above a certain threshold, the lake stopped recovering. Instead it started getting attracted to a new state with different population numbers as compared to the lake's initial state. This state consists of a new set of stable populations. As we've observed with other complex systems, if small numbers of bass are introduced, the ecosystem could easily recover (falling back to its initial attractor). As the bass number increased, the lake started moving closer towards the critical line – becoming more resistant to recovery. Once the critical line is crossed, the ecosystem falls into the domain of the new attractor – permanently changing the population sizes in the food chain to suit the new equilibrium. If we continue to meddle with the population numbers after this transition, the population would recover back to the new attractor state.

A group of Dutch ecologists researching the growth of trees in Africa, Australia and South America found that their development was highly influenced by attractors (Hirota et al., 2011). Due to the scale of study, the researcher conducted natural experiments – taking advantage of climate change and the way it was affecting trees. They found that tree growth responded directly to precipitation. When precipitation is sufficiently high, tree growth is influenced by the first attractor, forest zone. As precipitation crosses the critical line, the trees suddenly enter into the second attractor (savannah). If the trees are exposed to continuous lack of precipitation (perhaps due to lack of rainfall), they could still endure being in a savannah state – that is until the next critical line is crossed. They then enter into a treeless attractor state having only minimal undergrowth. These three main attractors were seen to drive tree growth in subtropical regions around the word.

More recently, attractors were also found to be behind the spread of disease epidemics (Dibble et al., 2016). These were the findings of Project AERO, led by ecologist John M. Drake from the Odum School of Ecology, University of Georgia. They discovered that there is a time delay between the outbreak of a disease (attractor 1) and the point where it becomes an epidemic (attractor 2). Again the two attractors are separated by a critical region – the tipping point. They defined the critical region as the point when infected individuals spread the disease to more than one individual. At that tipping point, the disease spreads uncontrollably becoming an epidemic (similar to the current Zika cases). The research team at Odum designed a computer model to identify the possible factors influencing the movement of the disease towards the critical region before it gets drawn by the 'epidemic attractor'. After numerous computer simulations, the team found that two main factors influenced the departure of the disease from the 'outbreak attractor' towards the 'epidemic attractor'. First being the rate at which the disease vector (i.e., bacteria, virus or protozoa) reproduces. The faster they multiply, the easier they get individuals infected which speeds the disease to the epidemic attractor. The second factor is the rate the disease spreads among the infected individuals. This factor is influenced by how the disease propagates – whether the individuals infect others via touch, sexual relations or other modes of transmission.

According to the team, the faster the disease reaches the critical region, the sooner it becomes an epidemic. This makes understanding the delay time between the two attractors crucial, as said by Christopher J. Dibble, one of the members in the research team – 'We know that there's some amount of delay between the theoretical point and the first major outbreak, and this effort was aimed in part at finding some likely values for that waiting time and in part to discern what drives it. The faster you approach that transition, the less waiting time you'll have. Once we understand this waiting time, we can learn from past epidemics. Using some of the modeling approaches that we've taken here we can extrapolate backward. If we had an epidemic in March, for instance, it's really tempting to assume that something must have happened in February or January to cause this population to change in a way that allowed it to experience this outbreak. But the problem is that the important change might have happened well before that. If we don't account for the delay between an actual outbreak and the point at which it was theoretically possible, we may be looking in the wrong place for the driver' (Gavrilles, 2016). The team at University of Georgia aims to develop an early warning system which tells if a disease is going to turn epidemic.

PATTERNS OF COMPLEXITY

Besides Schrödinger, the then newly introduced quantum theory was very much influenced by the ideas of Eugene Wigner. A chemical engineer by training, Wigner's knowledge of mathematics and physics were mostly self-taught. Born in Hungary, Wigner spent most of his adult life in the U.S. During the Second World War, Wigner was among the scientists who worked in the Manhattan Project developing the first atomic bomb. Observing the physics of systems with many particles (such as the uranium atom), Wigner went on to propose the foundations of a concept known today as 'universality'. Previously, we saw how complex systems with many sub-components peculiarly behave concealing multiple attractors (e.g., ecosystems, epidemics, forest growth). Like attractors, 'universality' is another feature which complex systems manifest. First observed in uranium atoms, systems showing universality have been popping-up in various places ever since. So what's universality? Universality is *a group of certain features* which recur in different complex systems (e.g., stock markets, ecosystems or the internet). Universality is used to determine the location of the critical line in a complex system (Nadal et al., 2011). Consider the Peter Lake experiment; universality could be used to calculate the maximum limit of the number of bass before the system tips into the next attractor. Take another example, the uranium atom – a complex atomic giant with 146 neutrons, 92 electrons and 92 protons. Using universality, we could determine the minimal mass of Uranium required to initiate a nuclear reaction. This feature is 'universally' observed although the systems in question are driven by different underlying mechanisms e.g., stock markets are driven by the trading of stocks, the internet works by rapid information transfer while ecosystems operate due to the network of food chains. Take note of the two criteria essential for universality; (i) the system has to have many components or parts (ii) it only occurs in complex systems – where its components/parts are interlinked in a complicated way.

Waiting for a bus in Mexico, the Czech physicist, Peter Šeba, observed universality occurring in bus systems. Šeba observed some men passing information to drivers about bus timings – specifically the time the previous bus left the station (Wolchover, 2014b). This info is sold on pieces of papers to bus drivers. The collapse of centralized bus system in Mexico gave birth to this new occupation called 'bus spying'. Bus spies sit at bus stops and record the comings and goings of buses. Bus drivers maximize their profit by taking as much passengers as possible at each stop. If the drivers drive too fast (they would be too close to the bus ahead), which would have taken all the passengers at the bus stop leaving no passengers for the driver behind. On the other hand, if they're too slow, the passenger build-up would be too much at each stop, making the bus system inefficient – making them lose customers. This is where bus spies come in. Using the info sold to the drivers by the bus spies, drivers could adjust their speeds accordingly to optimize passenger pick-ups. This spaces-out the buses at the right intervals picking up passengers accordingly as they accumulate at the stops. Šeba asked the bus spies if he could take a look at their records. At first they refused him, afraid that he was the police or some kind of enforcement official. When he convinced them he was simply a 'crazy scientist', they shared their records (naturally!). He saw that the trends in the bus spy records were similar those seen in certain types of quantum systems. A closer look at the records showed that these ad hoc bus systems shared many features with atomic systems, like uranium atoms. Šeba recognized that they were looking at a type of 'universal' feature occurring across different systems.

After Šeba, universality was also seen to occur in the complicated structure of sea ice. Universality has even cropped up in climate studies – which is not surprising since the climate is a complex network of many interconnected subsystems (e.g., regional weather patterns, storms and wind movement). The interactions of these subsystems make a complex network manifesting characteristics like attractors or universality. We know that attractors are the system's equilibrium points – there the system is at its most stable (or relaxed) state with minimal forces acting on it. Systems tend to move towards these attractors.

Although universality often emerges in complex systems, scientists don't exactly know why it's there – only that it gives us some idea about the critical region of the system. Many complex systems around us have critical regions. Take water for instance. If we apply heat to water at room temperature (27°C), it remains in liquid form until its temperature reaches its boiling point, 100°C. Then it changes phase becoming gas. The critical line for water is 0°C where it becomes ice (solid) and 100°C where is evaporates becoming gaseous. But is water a complex system? Yes it is. It may not seem so at first glance, but at microscopic scales we could observe that it is comprised of little molecules jiggling away interacting with one another. No matter how much or how fast heat is supplied to liquid water, it will not evaporate (or change phase) until it reaches its critical point of 100°C. The same applies for cooling; it has to get to 0°C to form ice. During phase change, say from liquid to vapor, the water molecules would have to acquire enough energy (from heating) to break its bonds with other water molecules and free itself becoming vapor. The reverse happens if it changes phase into a solid (ice) via cooling. Phase change is influenced by its internal molecular organization – which is affected by the level of energy possessed by the molecules. As you may have deduced by now, universality is present in water. Similarly, other compounds like petrol or nitrogen which undergo 'phase change' also manifest universality.

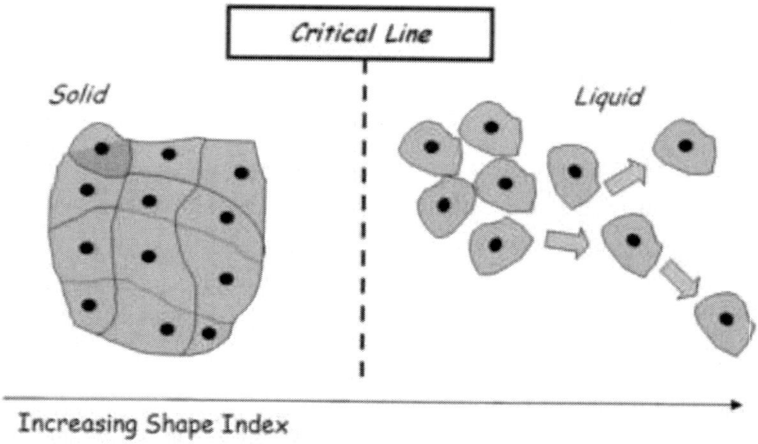

Figure 4.5. Cellular Phase Change.

Complex systems with many components tend to have interesting characteristics. Even if each of component seems to behave in a somewhat random manner, the system's collective behavior does show some distinctive patterns. Take cancer cells for instance. Peter Friedl, a biomedical researcher working at McGill University in Montreal, Canada found striking similarities between cancer cells and phase change in substances (Popkin, 2016). Cancer cells are known to metastasize – where they leave their original tumor and use the blood stream (or the lymph system) to move to other parts of the body. In the early nineties, Friedl used a microscope to take a closer look at cancer cells in motion. He observed the cells moving in clusters. Scientists today know that 90 percent of patient deaths from cancer result from metastasizing cancer cells (Spano et al.,2012). Almost two decades after Friedl's observation, the bioengineering professor at Harvard University, Jeffrey Fredberg was seeing similar effects in his research – moving clusters of lung cancer cells. In light of this results, Fredberg went on to propose a sort of 'jamming' index to measure the level of clustering in the cancer cells (Sadati et al., 2013). This is because the jamming of these cells seem to be linked to the potential for it become mobile (metastasize). Fredberg suggested that cells which are well jammed/packed would remain locally held in one place. On the other hand if unjamming happens, similar to the phase change in water; then the cells begin to metastasize becoming dangerous. This line of thought connects the propagation of cancer to the physics of phase change. Bear in mind that unlike non-living materials, cells are not consistent in their type (being not homogeneous). Cells across the body differ depending on their function. According to Fredberg they nevertheless share some properties with the physics of materials. His research suggests that even diseases like asthma could be caused by cellular jamming (Trepat et al., 2009). Inspired by Fredberg's work, Lisa Manning, a physicist at Syracuse University set out to determine this mysterious jamming factor – which she calls the 'shape index'. This index gives a crisp value representing how much the tissue is jammed (Figure 4.5).

Instead of pointing to the usual culprit; genetics, this angle of research considers the mechanical aspects of cancer. We may not be able crack complicated diseases like cancer by having a one-dimensional view on the problem. An interdisciplinary framework would shed more light covering various aspects of the disease, as stated by Fredberg – "People had always thought that the mechanical implications were at the most downstream end of the causal cascade, and at the most upstream end are genetic and epigenetic factors. Then people discovered that physical forces and mechanical events actually can be upstream of genetic events — that cells are very aware of their mechanical microenvironments." Progressing these ideas, the physicist Josef Käs conducted cutting edge experiments to study in detail the jamming of tumor cells (Fritsch et al., 2010). Käs thinks that if a whole tumor is jammed, it would become localized preventing it from metastasizing – making it benign. He believes that such a method may delay or stop the tumor's deadly progression. The jamming strategy is more of an engineering way of curing cancer. It doesn't answer questions like where tumors originate and how. It simply provides a possible and more efficient mode of halting it compared to our current methods of treatment.

THE GAMING EVOLUTION

Earlier in this chapter, we saw how Guy Theraulaz and his team from the University of Toulouse employed computer models to run virtual experiments to study the behavior of insects. Similarly computers have also been used to deepen our understanding of evolution. Much like the natural world, the first signs of simulated life 'emerged' out of computers in the late sixties. John Conway, a British mathematician working at Cambridge, devised a computer game he simply calls 'Life' (or 'Game of Life'). Unlike current advanced computer games, there is not much interaction with the player. The player only sets the initial condition of the game. The game is then let to progress on its own. The setting itself is pretty simple – a two-dimensional grid of black and white colored cells (see Figure 4.6). Then Conway placed a series of simple rules like; if the cell to the left is black, flip the current cell's color, otherwise maintain current color. Another example is; flip current cell's color to the color of the cell two cells above it. Using these simple rules the grid is then 'iterated'. Meaning that, beginning with the initial state of the grid (State 0), the next state of the grid (State 1) is determined using the set of rules. Then the subsequent state (State 2) is determined from State 1 using the same rules. This way the grid is continuously iterated determining new states from previous ones. Conway was surprised to see that as the grid iterated, new patterns started to emerge. It was as though the grid was teeming with life forms, merging, interacting, moving around and even reproducing. There were cases where some patterns would grow, thrive and eventually die (disappear) as the grid is iterated.

Until today various interesting patterns popping out from the Game of Life are being identified (e.g., glider, block, and beehive). The type of stuff that emerges depends mostly on the initial state (State 0) of the game and the number of times the grid is iterated. Conway's game gave the idea that it may be possible to simulate life using computers; and most importantly by using just simple rules. Ancient Japanese believe that the game of Go is the microcosm of the universe – where the possibilities of game play is staggeringly high. Like Go, scientists started looking at Conway's game as the microcosm of the evolution of life.

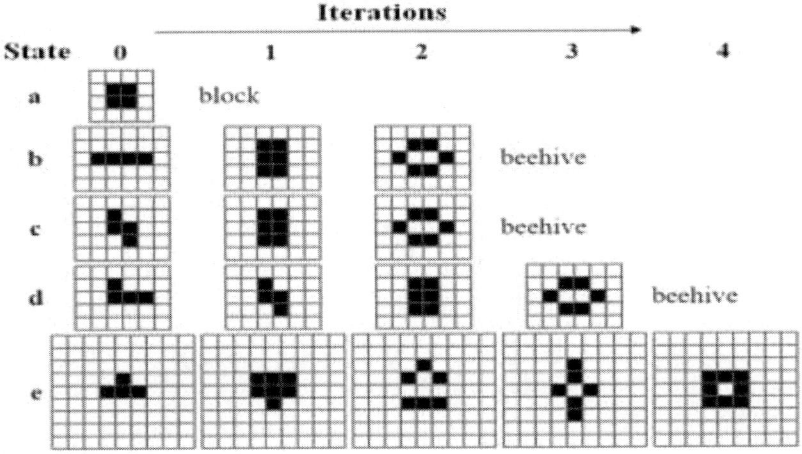

Figure 4.6. Patterns emerging from Conway's Game of Life (Gardner, 1970).

Most games require players to interact frequently to gain advantages over their opponents with the goal of winning in mind. Since evolution is all about competition, games are very useful models for thinking about animal interactions which result in their evolution. Biologists often use what mathematician's call the 'Hawk-Dove Game' (which a version of the famed Game Theory) to model animal interaction scenarios. Game Theory was founded by the mathematician and Nobel Laureate John Forber Nash – played by Russell Crowe in the film 'Beautiful Mind'. The Hawk-Dove game explains why animals don't constantly fight. There are two main strategies in this game; aggressive and peaceful. The hawks are associated with being aggressive while the doves are peaceful. The game involves hawks and doves going after resources like food to survive. During confrontation, the aggressive strategy of the hawk causes the dove to back down, letting the hawk have the resource. This way hawks and doves avoid conflict. At first glance, it would seem that the repeating of this scenario would cause the doves to die out of hunger. But could hawks could go on fighting without wearing themselves down? Being constantly in combat, the hawk even neglects spending time nurturing its offspring. Thus, battling away could become detrimental for the hawk in the long run. According to game theory, there exists a compromise between the amount of fighting and the time spent in the nest with the offspring. Since the hawk doesn't keep battling, the dove gets its share of the food. This way the resource is shared and both birds fare well in the long run. Recent research shows that the hawk-dove game theory could be used to predict various animal populations – for instance Australia's Gouldian Finch (Kokko et al., 2006; Kokko et al., 2014; Salleh, 2014).

Using the hawk-dove model, scientists have now studied how population sizes of bald eagles are affected by their fighting behavior (Hansen, 1986; Garcelon, 1990). Besides competition over resources, the hawk-dove system has been applied to model competition for mating. In the case of the bowl-and-doily spider, the males engage in combat to mate with the females. The reward here is the ability to reproduce. However, they must risk serious injury in this pursuit – which could possibly result in death. Additionally, the female spiders have different levels of desirability by the males (based on the number of eggs their capable of fertilizing). Accounting for all this consideration, the interactions among bowl-and-doily spider could be determined using hawk-dove game theory (Austad, 1983).

EVOLUTION *IN SILICO*

Treating nature like a game spurred scientists to use computer models to simulate evolution. The beauty of computer models is that one could perform millions of virtual experiments simultaneously and get quick results. Most evolution simulators are built using the concept of genetic algorithms; pioneered by engineer and computer scientist, John Holland. Working on the question of how life forms adapt in nature, Holland developed genetic algorithms (Holland, 1984). Algorithms are guidelines or rules used to create computer programs. With the right algorithm in hand, computer programs doing specific tasks could be simply and effectively written. Therefore, using genetic algorithms, Holland created programs capable of evolving and adapting in a computer. This programs were used for simulating real natural systems. Today such software is known as evolutionary programs. The simplicity of Holland's genetic algorithm made it easy to implement in various fields.

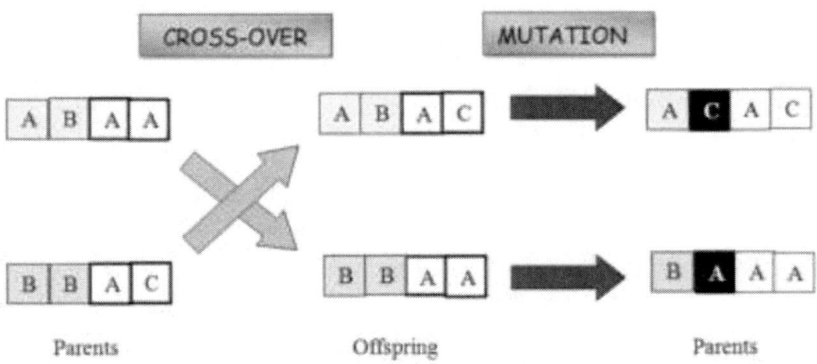

Figure 4.7. Artificial Evolution and Adaptation using Genetic Algorithms.

Evolutionary programs were thus widely developed and used across various disciplines (besides biology). Virtual evolution is achieved using genetic algorithms via two primary mechanisms; cross-over and mutation. This is how most evolutionary programs work. Say we want to simulate an ecosystem – we'll use the familiar Peter Lake example. First we virtually create the inhabitants: bass, minnows, fleas and algae. Assuming we are only interested in the evolution of bass – so we want to only simulate this evolution. Using the genetic algorithm, we first define the traits of the bass in an individual's genome. The three traits being teeth sharpness, fin length and body strength. Initially we randomly assign the traits to the bass population. Then the fishes are crossed-over or 'mated'. During this process the traits from each individual fish are 'crossed' with another. The offspring resulting from the crossing of two parent fishes are unique – consisting partial traits from each parent. Based on the environment, the offspring's genes are then mutated creating further variation in its genome. The cycle then repeats where the offspring plays the role of parents and creates more fishes. Over many cycles or 'iterations', adaptation starts to emerge and the virtual environment created using genetic algorithms start behaving like the real thing. This process is depicted in Figure 4.7 where the traits are symbolized using three alphabets (A, B and C). In Figure 4.7, two traits are crossed–over from parents to offspring while the purple traits indicate mutation in the offspring's gene code.

Now that we understand genetic algorithms, let us take a look at how these algorithms have helped biologists in their work. Flatworms have been intensively studied by scientists for over a hundred years.

Among the initial pioneers of flatworm research is Thomas Hunt Morgan (Graber, 2016). Morgan mostly studied genetics and evolution. Nevertheless towards the end of the 19th century he began spiking interest in the area of regeneration. He noticed the strange regeneration ability of certain animals, like flatworms, tadpoles and some types of fish. In his studies he performed careful dissections on flatworm and observed how they regenerated every time – growing a complete new body. Fascinated by this he set out to determine the number of times he could keep slicing a flatworm until it stops regenerating. Morgan found that this happens when you slice a flatworm into two at its midsection and go on doing it for 279 times. At the 1/279th cut, flatworms will fail to regenerate and until that number is achieved the flatworm will keep regenerating creating new individuals with every slice (Figure 4.8).

Morgan's investigations got biologists curious as to what caused such regeneration in animals. After Morgan, researchers busied themselves trying to figure out the regeneration puzzle. Their modus operandi was twofold: genes or chemicals. The genetic approach involves attempting to knock-out or activate certain genes and see what happens e.g., if the flatworm grows another head or its tail disappears. The chemical approach is; observe changes in the flatworm by exposing it to various chemicals or by altering its body's own chemistry. These sorts of experiments became popular among biologists studying regeneration. Thanks to all the tinkering, a large wealth of data has been accumulated since the time of Morgan. Even with such vast amounts of data, biologists still cannot seem to understand how it all comes together to answer the very first question – how do flatworms regenerate?

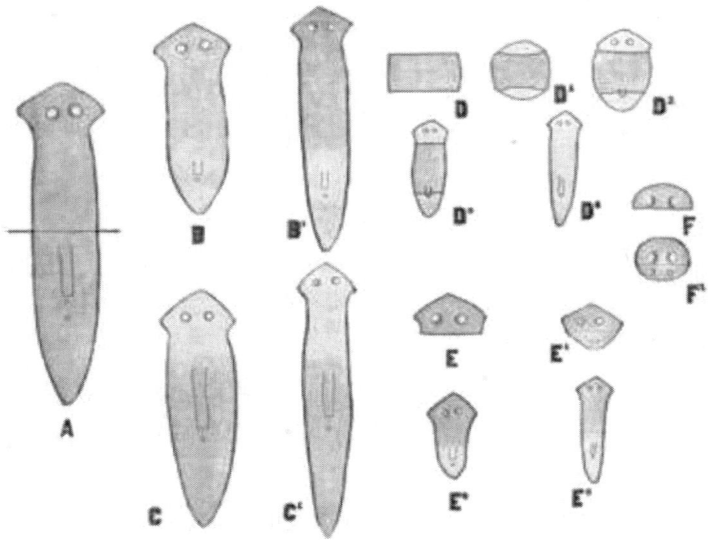

Figure 4.8. Thomas Hunt Morgan's depiction of flatworm regeneration from his book entitled, *Regeneration* (Hunt, 1901).

Recently, Michael Levin, a scientist working at Tufts University, Boston tried to tackle the flatworm puzzle using an unconventional approach; genetic algorithms. Levin wanted to develop a computerized worm which simulates the regeneration behavior of real-life flatworms. To create this virtual worm, Levin accumulated all the flatworm tinkering data he could get his hands on. He then used genetic algorithm-based programs (evolutionary programs) to learn and develop a model based on this library of data. These evolutionary programs develop and improve millions of flatworm models simultaneously – consolidating all the experimental findings for the past 100 years or so.

Levin's approach to the flatworm puzzle is what physicists (and engineers) call the inverse problem method. The inverse problem method is basically 'reverse engineering'. Usually physicists tackle problems by developing models using mathematics and physics. Being consistent with all our past knowledge, this model solves the problem at hand. The inverse problem method solves the problem the other way around – where vast amounts of data is collected and used to develop a model. Models developed using this method does not need to be consistent with past scientific knowledge and theories. Using powerful computers it simply solves the problem. Levin's virtual flatworm strategy is definitely an inverse problem method since it 'evolves' the model to fit the vast library of data. As explained by Johannes Jaeger, a geneticist at the Korand Lorenz Institute, Austria – "This approach—reverse engineering—it's like a Russian spy movie. You have some kind of gadget, like in the Cold War, some kind of Russian technology. You don't know what it does. Here, you have an organism, you don't know how it works, and you're trying to infer the networks that make the pattern that you see in animals."

Levin together with his colleague Daniel Lobo (a computer scientist) set out to develop a virtual flatworm model which learns from the library of data. They quickly realized that regular computers in their laboratories could not handle the processing work. Therefore, they rented time on the University of Texas's supercomputer; called Stampede. Stampede is capable of performing quadrillion mathematical operations per second. A quadrillion is 1 with 15 zeroes after it (10^{15}). This makes Stampede a computational behemoth! Evolving models using Stampede, the 100th generation of models showed promising features – they started emulating real flatworms. Levin and Lobo checked this through by performing virtual incisions and comparing the results with the ones done in real-life. Stampede took 26,727 generations of models and about 42 hours of number crunching to churn out a single viable model. But it was definitely worth the effort since the final model could explain 1000 physical flatworm experiments. Their virtual flatworm model even started showing some unknown features never seen before in lab experiments. The virtual flatworm hinted at an unknown protein which seem to be playing an integral part in flatworm regeneration. Lobo and Levin's strategy using genetic algorithms was surely a breadth of fresh air after over a hundred years of research into regeneration. Most importantly, their approach opens the door to a new method of biological exploration – the concept of experimenting with virtual organisms. This way scientist can by-pass a whole lot financial and ethical issues that comes from running experiments with real organism. Besides, with the abundance of computational power provided by machines likes Stampede, biologists can now harness it to aid in their research. Approaches such as genetic algorithms have taken-off successfully in areas like engineering and computer science. Today genetic algorithms have improved engineering design and operations across various applications (Ganesan et al., 2016).

RISE OF MACHINE INTELLIGENCE

Until recently, computers could not perform tasks that require learning, although we could program it to perform complex tasks. For instance, "Deep Blue' the chess program which played again Gary Kasparov was certainly complex. However, such computers need to be programmed by savant-level chess masters and computer scientists to anticipate the opponent's moves and move pieces accordingly. Programs like 'Deep Blue' cannot learn and move autonomously during gameplay. At that time, we were not able to program it in such a way to have innate learning capability. In the last decade, computer scientists designed an algorithm which could provide computers with this 'innate learning ability'. This algorithm emulates the human brain – especially in the way it learns and stores data. The algorithm is used to design an 'artificial neural network'. It consists of multiple individual neurons connected with one another to make up a complete neural 'net' – just like in real brains. In this context the whole thing; neurons and all are completely virtual. This form of artificial learning is called 'machine learning'.

So how does it learn? Easy, first you expose the program to the data or environment it needs to learn. This process is called training. Once it's 'trained', you can use it to predict or anticipate data – based on what it has learned. The neural network wouldn't be able to predict completely unfamiliar data – data it was not trained with. Its predictions are based on the best of its knowledge which is information it learned during training. Conventional neural network programs have three main layers; input and output neurons along with a hidden layer (refer to Figure 4.9). This is how the network learns: First the input and output layers of the neural network are exposed to data for it to learn. Here the input neurons receive data which causes the outcome which is preset at the output neurons. The neural network is trained with events where the outcomes and causes are clearly understood – where we know that certain data (input) causes some outcomes (represented by the output data). Then the network starts to iterate; repeatedly assigning the correct parameters to the hidden neurons. Once this is established, we can safely say that the neural network has learned. Now comes the second part; using what it has learned for prediction. Here the network is presented with new data at its input neurons. Using the parameters it established at its hidden neurons (via learning), it predicts the data at the output neurons. This is how an artificial neural network could be made to learn and provide us with useful predictions. The invention of artificial neural networks is certainly one of the hallmarks of artificial intelligence. Having computers which can learn opens up many new possibilities in many areas of scientific inquiry. Developments in computer intelligence directly influence our understanding of genetics.

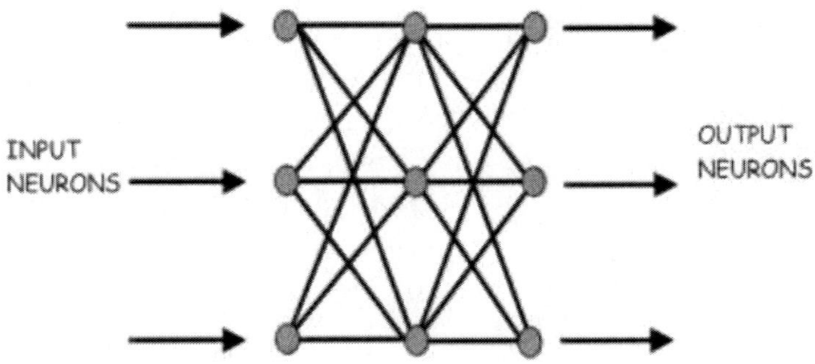

Figure 4.9. Neural Network Diagram.

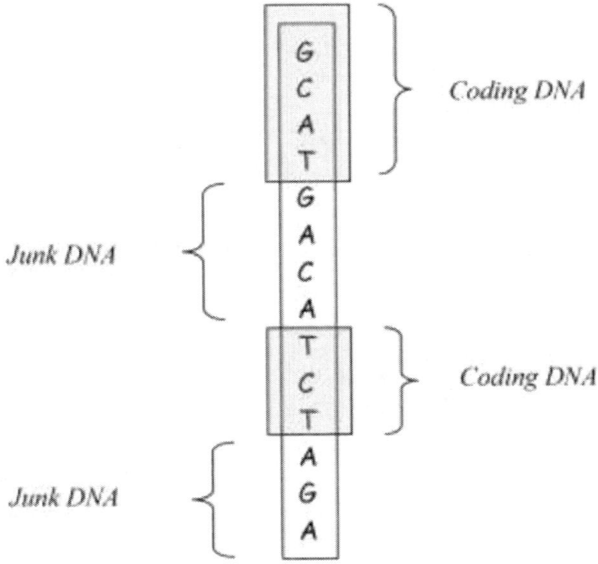

Figure 4.10. An example of a single line of genetic code for developing a specific protein.

In Chapter 1, we saw how the DNA provides the cell with codes needed to build proteins; which do most of the cellular 'legwork' (see Figure 1.4). Nevertheless, only a small part of the DNA is used to directly build proteins (about 1%). This part of the DNA is the 'coding DNA'. Strangely, most part of the DNA is non-coding, or better known as 'junk DNA'. Junk DNA does other stuff like deciding when to start a new code or how to arrange the genes so that they are meaningful enough for manufacturing proteins. Bear in mind that meaningful instructions from the DNA are often presented not by a single gene, but by a network of genes. Imagine the DNA as a string of words on a page, say describing how to ride a jet ski. For the description to be accurate and meaningful, the text (coding DNA) would have to be arranged in a coherent manner – with full-stops, paragraphs, spaces and bullet-points, all in the right places. Additionally it needs correct grammar; so that the message gets across effectively. These rules and conventions are regulated and put in-place by the junk DNA – which is why it's known as 'non-coding DNA'. Junk DNA does not provide the cell with the instructions to build proteins, it only provides the rules of how the instructions are written. Figure 4.10 shows how junk DNA and coding DNA are positioned in a single line of genetic code.

Let's say the gene code in Figure 4.10 presents the instructions for developing a specific protein. Then the way it's written is influenced by the codes and the positioning of the junk DNA. This influences how the coding DNA is read which in turn determines what type of protein is made. The coding DNA has been explored widely by geneticist. However, much of junk DNA (and how it regulates the coding DNA) still remains a mystery. In 2010, Brendan Frey and his team working at the University of Toronto managed to identify junk DNA in a mouse genome (Barash et al., 2010; Xiong et al., 2015). Following this achievement they continued to collect loads of data from the human genome. Frey and his team collected two types of data:

1. Variations in the junk DNA (input).
2. Data detailing the effects of junk DNA variation on coding DNA (output).

Most of the data were taken from the genome of people suffering from afflictions like muscular atrophy and colorectal cancer. The data showed how variations in the junk DNA codes could lead to the manufacture of wrong proteins which results in those diseases. This way they correlated the effects of junk DNA variations to the coding DNA. As simple as this may sound, we are talking about linking-up large sets of complex data. The team attempted to deal with this using a machine learning approach called 'deep learning' – similar to a neural network. Frey and his team trained the network to obtain a predictive computer model. This model predicts the type of proteins (output) which are built given a variety of junk DNA code (input). Thus it could predict the resulting disease caused by variation in junk DNA.

Frey describes his approach to cracking the junk DNA enigma as follows – "It's one thing to have the book — the big question is how you read the book. When it comes to the genome, context is important, just like in English. 'Cat' means different things whether we're talking about pets or construction equipment. In the same way, how the cell interprets a set of splicing instructions depends on other nearby instructions. A string of DNA that means 'make lots of component X' might mean 'don't make component X' when it's sitting near a second set of instructions. Whether a sequence has an effect depends on whether another sequence has an effect. Without understanding that, it's hard to predict how a pattern will affect splicing" (Singer, 2014b). Using models like the one developed by Frey's team, doctors could watch out for dangerous junk DNA variations in patients – gauging if they are predisposed to diseases like certain types of cancers. Besides, using machine learning we could also learn of undiscovered diseases which could potentially result from junk DNA variations – sort of a disease forecasting system.

Another affliction that has been recently linked to genetics is autism. Although scientists know there exists a link, the autism genetic markers vary from one individual to another. This means that the genetic variations between autism patients significantly differ – making it difficult to identify and generalize. In many cases, the autism genes of patients need to be cross-checked with other family members to make sure that the suspected genes are not simply a common genetic variation present in the family. Narrowing-in on the gene culprits can be very tricky. What we'll need is a system that could work through vast amounts of data and syphon out autism genes.

Research teams from Princeton and the Simons Foundation thought that this problem was tailored for machine intelligence (Krishnan et al., 2016). A quick refresher: genes do not operate individually, they need to be regulated and written with correct and meaningful 'grammar'. And this is done thanks to junk DNA. A cluster of genes written correctly for functioning are called 'genetic networks'. If any part of the network malfunctions (or is absent), then the functionality of the gene network may be distorted, resulting in disorders like autism. Therefore, scientists at Princeton used a neural network program to comb through all this data and learn their anomalies. The program then cross checks new genetic data for similarities. This way they zoomed-in their search on the potential genetic network failures that may result in autism. Olga Troyanskaya, one of the Princeton scientist explains, "We don't just say that [a gene] has a 90 percent chance of being autism-associated, because we have the network (genetic) we can actually say, 'This is how it's connected to autism'. If something goes wrong, like you don't have one of those genes to bind together, then they're not going to go on and bind the third gene, and now you broke this little circuit "(Gershgorn, 2016) The machine learning approach employed by the Princeton Research team has not identified the genetic culprits causing autism. Nevertheless, their approach is making serious strides towards getting to the crux of the matter when it comes to the genetics of autism.

Helping with Dementia: Parkinson's, Alzheimer's and ALS

Machine learning has also been applied to develop strategies in preventive medicine. A group of Finnish doctors and engineers recently used machine learning to design an index to measure the risk of dementia in older people (Solomon, 2016; Pekkala et al., 2017). We now know that lifestyle is among the key factors contributing dementia. But we still do not have an effective tool which predicts the initial onset of dementia; early-stage prediction. Recent research into dementia in Finland involved the collection of huge piles of data. To help efficiently store, manage and access all this data, an online data storage system named Brain Health Registries was developed. Just like to the virtual flatworm program mentioned in the last section, Finnish scientists used machine intelligence to learn and handle large amounts of data they pooled from the Brain Health Registries. Their data mostly consisted of Finnish individuals aged from 65 to 79 years. The subjects underwent various memory and cognitive examinations. Using data from 10 years back, the machine learning program was trained to identify key factors for predicting the development of dementia. The program identified the following factors:

1. Cognition – mental capability.
2. Subject's Age.
3. Vascular conditions – for subjects having clogged arteries, blood pressure issues or other blood flow-related health problems).
4. Frequency and intensity of memory complaints.
5. Existence of a gene called E (APOE).

Using these factors, the resulting dementia index managed to successfully predict subjects currently developing dementia using data from 10 years back. The Finnish researchers cautioned that the index does not serve as a comprehensive dementia prediction method. It is rather a tool which may aid in detecting dementia in the initial stages – suitable to be used as a support for clinical decision-making.

Machine learning is also being considered as a tool to directly diagnose patients with Alzheimer's disease. As with other types of dementia, physicians have no way to diagnose patients having early stages of Alzheimer's. Changes in brain functions are usually observed using imaging technology like MRI (Magnetic Resonance Imaging). Although certain changes occur in the brains of Alzheimer's patients, until now radiologists and physicians could not pick-up similarities between these changes. It they can do this, then they can correlate these changes to early-stage Alzheimer's disease.

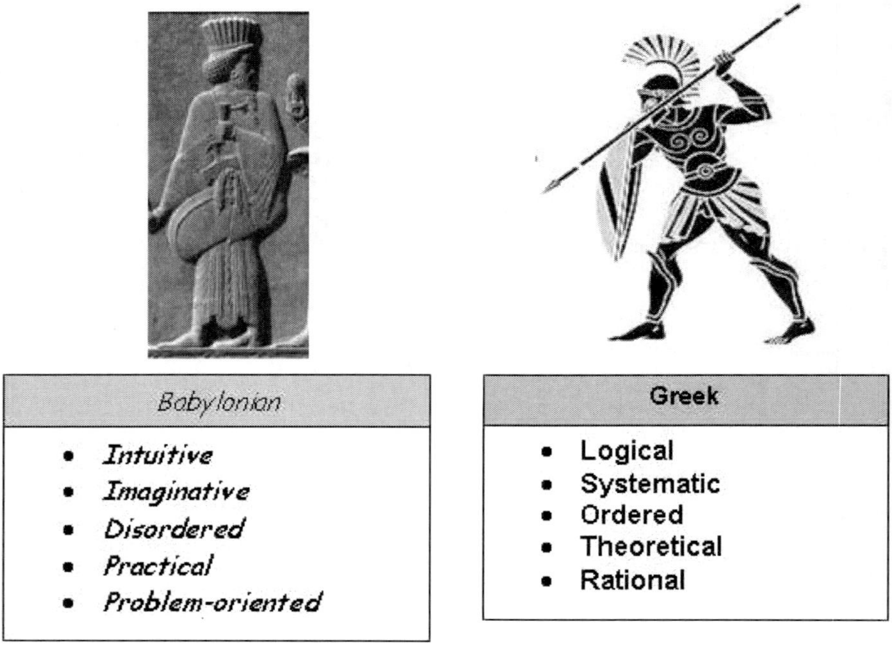

Figure 4.11. Thinking Styles – Babylonian versus Greek.

A group of Dutch research scientists led by Alle Meije Wink, from VU University Medical Center, Amsterdam used a machine intelligence program to tackle the problem (Casey, 2016). They trained their machine intelligence program to learn data from MRI scans of brains with previous patients who had early-stage Alzheimer's disease. The program learned all the patterns in the brain map from the scans and identified some similarities. They used an open-access machine learning software, Pattern Recognition for Neuroimaging Toolbox (Pronto). Using this software, Wink's research team managed to predict early-stage Alzheimer's in patients with a staggering accuracy of 82 to 90% (Collij et al., 2016).

Two tech companies, Origent Data Sciences and Cytokinetics started collaborating to develop a computer model for detecting ALS progression in patients (Semedo, 2016). Besides early-stage detection, even determining the progress of certain types of neurological conditions, (like ALS) have been difficult. Among the reasons for this; is the individuality of the brain. Since every brain is complex and highly individual, the way

ALS affects it is unique. The same goes for the way it progresses. This makes systematically measuring progress for individual patients extremely difficult. The two companies employed something called the Gradient Boosting Machine (GBM) algorithm which is a machine learning–based program. As mentioned in the previous chapter, ALS is neurodegenerative disease that affects muscle function across the body. It therefore weakens the muscles controlling the lungs – decaying the functionality of the lung as the disease progresses. The developed machine-learning program determines the level of ALS progression by measuring the degree of lung functionality. The program was used to learn from lung function data from 8500 patients. It successfully churned out 22 baseline characteristics which can be used to determine ALS progression in different patients. But as with most machine learning software, the more data it's exposed to, the more it can learn. And the more it learns, the better its predictive abilities. Having a strong machine learning program is not very useful without access to large amounts of data for it to learn from.

Speech pattern is also another avenue for spotting dementia. An American start-up, Canary Speech worked on a project called Echo – a research effort to develop a successful technology for identifying dementia. The speech recognition software parses through the words being said and looks for cues signifying dementia. Their software was set to learn vast amounts of data consisting mainly communication recordings between clinicians and their patients.

Thinking about machine learning and other A.I. approaches, we cannot help but coming back to the inverse problem concept we visited above. Let us digress into a bit of philosophy to get a clearer picture of the inverse problem method and how it relates to machine learning. The style of thinking the inverse problem method offers is not a new one. Leonard Mlodinow, a popular science author captured some of these ideas in a book he wrote about his conversations with the great physicist, Richard Feynman (Mlodinow, 2004). Among the main ideas presented there are the two opposing (or complementing, depending on how you look at it) philosophies developed by the Babylonians and the Greeks. According to Feynman, the foundation of western civilization were laid down by the

Babylonians. They obtained the great understanding of geometry and mathematics. Nevertheless we remember the Greeks as the founders of mathematics. This is because the Babylonians were masters in using the inverse problem method. They focused on solving the problem at hand. The Greeks on the contrary focused on the method of solving the problem. They were concerned if the method obeyed a set of laws and whether the method could consolidate with other methods, generalizing the approach. In short, the Greeks were suckers for 'systematization' while the Babylonians only cared about solving problems. Therefore, the Greeks could be thought of as rational, logical and theoretical thinkers; while the Babylonians we more intuitive, imaginative and practical (see Figure 4.11).

Feynman himself considered himself as a Babylonian-style physicist. The inverse problem method has existed since the time of Babylonians. This method is making a strong comeback in this age of data and artificial intelligence. Confused by an avalanche of various data, reasonable systems become increasingly difficult to establish; e.g., brain scan data for Alzheimer's and lung function data for identifying the progression of ALS. Hence, the best way to approach these types of confusing and complex scenarios is to focus on the problem at hand, Babylonian style. And the best way to go Babylonian is by using the powerful techniques offered by artificial intelligence; machine learning.

FEELING ELECTRIC

Since the discovery of the DNA along with the rise of genetics, biology and medicine has flourished and attained greater heights as never before. The understanding of genes somewhat unlocked for us the governing codes of the cell. This deeper understanding of life inspired many scientists to leave other interesting areas of research and turn their focus on genetics. Nevertheless people like Michael Levin persistently explore biology using various perspectives other than genetics. We are talking about the same person we discussed previously, Levin from Tufts University, Boston; the man who approached the flatworm puzzle using machine learning. When not tinkering with machine learning, Levin has a go at solving the electrical side of living things. The problem he was looking at still has something to do with regeneration (like the flatworm puzzle). The question this time was; what natural mechanism organizes a bunch of cells to form certain parts of the body? For instance, how does the body organize its cells to form a hand, a heart or any other organ? Since the genetic revolution, we have thought that all this could be linked to genes. This notion has been recently challenged. Scientists like Levin have shown that organ development is very much influenced by 'bioelectricity' – electricity produced by the organism's own body.

In the 18th century, bioelectricity was known as 'animal electricity'. Observations were made on animals like electric eels and torpedo fish (electric ray). These animals were seen capable of delivering electric shocks when intimidated or merely touched. The Italian doctor, Luigi Galvani was among the first to investigate this idea of animal electricity.

Working at the University of Bologna, Galvani found that electricity generated by a rotating electrical generator could cause contractions in the muscles of the legs of frogs (often covered in biology classes to simultaneously educate and gross out students. Figure 4.12 depicts the charging of an electrode with an electrical generator known as a 'Leyden Jar'. Then the electrode is made contact with the frog legs making the legs twitch. Extending his experiments, Galvani tried to connect frog legs to two different types of metals, instead of an electrical generator. Doing this he found that the legs similarly twitched, leading him to conclude that the tissue in the frog legs must be generating electricity per se – animal electricity. Upon reaching these results, Galvani published his findings to the scientific community.

Inspired by Galvani, another Italian scientist, Alessandro Volta repeated the frog-leg experiments and reached the same results (Sabbatini, 1998). The unit for electrical potential, Volt is named after Alessandro Volta for his groundbreaking research into electricity. Volta was the first to build a battery to store electricity. As for Galvani, his name lives on today in much scientific terminology. For instance, the Galvanometer is an instrument employed to detect electric current while Galvanic Corrosion is a type of aggressive form of corrosion occurring when two different types of metals come into electric contact. Muscular convulsion resulted from electrical stimulation is today called Galvanism.

Among the modern day champions of bioelectricity is Michael Levin. Levin and his team managed to alter the growth of many organisms without using genetics. Employing electricity, he managed to grow flatworms with two heads and frogs with six legs among other oddities. According to Levin, just as there is a genetic code, there exists a bioelectric code. He says, ‘Once we know how anatomy is encoded, we will be able to make shapes on demand’ (Bittel, 2017; Ledford, 2015). Levin’s technique involves varying the electrical gradient between the organism’s cells. By doing this he could somewhat control the types of cells growing at certain regions. So how does he do this? Since the 18th century, we have known that most cells grow and function by using electricity. Electrical signals are carried to and from the cells via specific channels. These channels could be thought of as organic ‘wires’. Electrical signals are then transferred in the form of ions (charge-carrying molecules). Levin and his research team control the electrical gradient between cells by adding or blocking the organic wires. Blocking these wires reduces the electrical gradient while adding them does the reverse. There are two ways to do this:

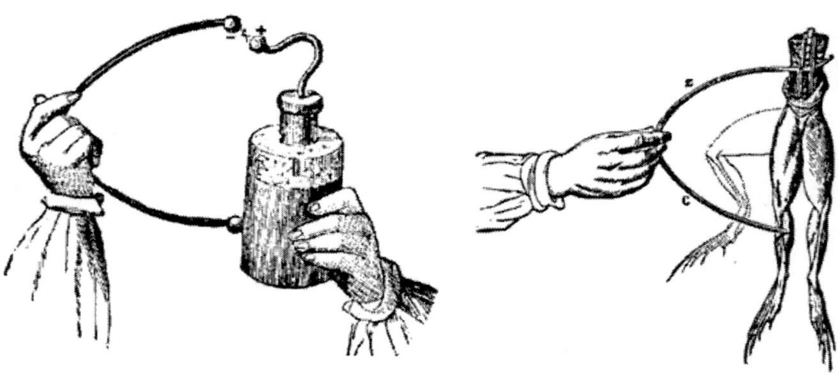

Figure 4.12. Galvani’s original diagram depicting the charging of the electrode and applying it on frog legs (Krueger, 1746; Wells, 1872).

1. Using an external electrical source to enhance or reduce the electrical signals among cells.
2. Drug Therapy – to chemically block or add ion channels (organic wires).

The Tufts University research group then proceeded to test their ideas on frogs (Pai et al., 2015). They manipulated the electrical gradient in frog embryos. This way they managed to promote brain growth in the frogs – developing extra brain tissue. Levin's team may have found a way to regenerate damaged brain tissue resulting from brain injuries and birth defects.

Levin's team picked frogs for their experiments because tadpoles have long been known to regenerate. If we sever a tadpole's legs or tail, it will easily grow it back – complete with all the blood vessels, tissue and even the spinal cord. Many small animals have freakishly remarkable regeneration capability. Most salamanders could regrow their tails, jaws, limbs, intestines and even their eyes and hearts (to a certain extent). Although regeneration is uncommon in large animals, there has been instances when it happens. One such case is in male deer. These deer have been known to grow back their broken antlers. As with the tadpoles, the regenerated antlers are complete and are no different from the originals. Even humans have some regenerative features. The most regenerative organ in the human body is the liver. The liver repeatedly recuperate after the most common damage – large consumptions of alcohol. Children below the age of seven have been known to regrow their fingertips after amputation. There have also been rare cases when patients who lost kidneys have been found to regrow the organ after some years (Ledford, 2007).

LIFECYCLE OF IDEAS

The idea of the bioelectric code is certainly a paradigm shift from the conventional genetics-based strategies. The development of ideas in science often has its own dynamics. Thanks to peer-review and other rigorous standards, scientific ideas go through a version of evolution. Ideas which stand scrutiny tend to live on while ideas which don't eventually breakdown and become forgotten. Strong ideas tend to get cited by other scientists furthering the initial work. The more citation the work gains, the longer its lifespan. This is because strong ideas attract more interest and inspire more researchers to explore further. This way the really great ideas get continuously referred to and so live on time immemorial. Standards aside, popularity also contributes to the lifecycle of ideas. It sometimes happens that, as popular ideas emerge, older concepts (albeit sometimes proven effective and useful) tend to slowly wither away along with the bad ones. This applies especially to ideas that challenge the status quo or threaten the dominant political powers.

The oldest story in the book is the ordeal of Galileo. Copernicus discovered that the planetary motion centered on the sun – the idea of the 'heliocentric model'. By the time of Galileo, the Catholic Church (Inquisition) dismissed the heliocentric model and declared it heresy. Championing the heliocentric model, Galileo was excommunicated by the Church and placed under house arrest. Another such case is Pasteur's germ theory. Until Louise Pasteur came about around 1861, the notion that microbes could causes illnesses was not widely accepted. In fact, many physicians refused to wash their hands when working between patients. They disregarded germ theory and even criticized it. It is often the case that really good ideas like germ theory just would not perish that easily although faced with heavy resistance.

Among those who acknowledged germ theory when it was still in its infancies was one John Snow. Snow was a British physician who investigated the outbreak of cholera. Inspired by germ theory, he believed that water contaminated by sewage was the cause of cholera. Cholera is an intestinal disease causing vomiting and diarrhea upon infection. Continuous infection could result in death. The cholera outbreak in London during the time of Snow killed thousands. Snow was a physician specializing in obstetrics – an area in medicine focusing on child birth and pregnancy. His interest spanned widely across diverse fields of medicine;

including epidemics. Scientists at the time thought that cholera was caused by breathing in infectious vapors and noxious pollutants present in the air. So when Snow published an article describing how microbes may be causing cholera to spread, he faced outright rejection from the scientific community. During that time, the public got their supply of water from the town wells and communal pumps. Many companies bottled their water directly from the Thames River. This water was used and sold in breweries and restaurants across London. The public disposed waste and sewage by channeling it directly into pits known as 'cesspools'. Using germ theory Snow developed a working hypothesis on how cholera was spreading. According to him the germs, from these cesspools were seeping into the town well contaminating the water supply. Drinking the contaminated water causes the cholera-causing germs to infect and spread widely among the public.

The London suburb at which Snow lived called Soho was hit hard by the cholera outbreak. This presented an opportunity for Snow to get some research done to prove his germ theory hypothesis. Snow immediately targeted the public water source at Soho; a street-pump in Broad Street. He then collected data from hospitals and other records regarding how many patients used the water supply from the street-pump. If his hypothesis is right, then those who worked and lived near the street-pump would be directly exposed to the germ. These people would be among the most likely to contract cholera. By mapping out the victims on a chart, he proved his hypothesis claiming the street-pump as the source of the outbreak. His data even indicated that restaurants and coffee shops serving glasses of water reported many cholera cases among their customers.

To add rigor to his method and rule out other sources of cholera, he investigated people who did not contract the disease. Snow wanted to know if these people drank from the street-pump. Then he investigated a prison which had no cases of cholera. He found that the prison had its own well and did not get its water supply from the public. Similarly a brewery which had its own well had no cholera cases. A factory which used the water from the street-pump at Broad Street had casualties. All these cases could be related to Snow's hypothesis using the street-pump as the source.

But one case eluded Snow; involving a niece and her aunt who both died from cholera but both women lived far from Soho. Talking to the women's son, Snow cracked the case. He found out that at one time, the aunt lived at Soho. She liked the taste of the water so much; she had the water from the pump bottled and brought to her. And when her niece came by visiting, she drank the same water. Both died the next day of cholera.

After all this research, Snow brought the matter to the municipal authorities. He convinced them that the street-pump was the source of cholera. Reluctantly the authorities removed the handle of the street-pump rendering it useless. The rate of cholera outbreak almost immediately reduced to a stop. Although Snow's strategy was effective in halting the disease propagation, government officials and the Board of Health adamantly refused to accept the theory. Additionally, the authorities refused to accept the notion that there may be a sewer line leaking into the main water supply. After all this, Snow continued to collect data maintaining his stand on the germ-based theory of cholera – which we know today to be a tested and verified fact (Tuthill, 2003; Friedman, 2014).

Herding Effect and Zombie Ideas

Scientific ideas like most other ideas are subject to the 'herding effect'. Herding effect happens when a group of followers get influenced by other followers forming a sizable 'herd'. Individuals forming the herd may not be rationally working towards a central goal but rather may just be simply mimicking the behavior of others in the herd. A prime example of such behavior could be seen in property purchase (mainly in real estate). Instead of digging up information about the soundness of the property chosen for investment, most people would opt to simply follow the direction of the herd. The larger the herd, the safer they feel – the more they invest in the chosen property. Herd behavior could be readily observed in financial markets; especially when it comes to investments.

Sometimes herd mentality also happens in science. The quality of scientific ideas is often instinctively ranked by certain dominant factors. First: prominence and popularity of the researcher who is propagating the idea. The more prominent the persona, the more credible the idea seems – even if it is completely theoretical, not validated or proven. The same goes for universities, institutes and research clusters. University and institutional ranking plays an integral role in piloting scientific ideas. Naturally, most media, science magazines and online articles only popularize scientific theories developed in high-ranking universities and prominent researchers. Besides, prominent researchers from well-known universities also get easily published in high-ranking journals – this in effect reinforces their status. One may argue that established universities and researchers act as a filter to the scientific community; only permitting those with sufficient scientific skill and aptitude to do science, thereby to produce high-quality work. Nevertheless such institutes and clusters also become avenues for herding behavior. Most scientists tend to cluster up with prominent researchers from high-ranking universities. They then simply follow the scientific ideas of the herd – without doubting the soundness of the herd's 'prominent' ideas. And the larger the research group or university, the more safer and confident they are with their ideas. Then they go ahead and publish their works in scientific journals edited and reviewed by researchers from their fraternity – rejecting any other idea (no matter how good) which opposes their scientific framework. Many may argue that the difficulty of getting great scientific ideas out adds rigor to science and filters out the shaky ideas (Poole, 2016). But this system may also kill-off some great ideas along with the lesser ones. It may also slow the progression of science due to the redundant obstacles good ideas would have to breach in order to be published and accepted in the scientific community. Such situations have been encountered before by researchers who unlocked complete areas in science such as John Snow, Louise Pasteur and Robert Koch.

The most interesting findings showing herd mentality was obtained by a research group at MIT (Muchnik et al., 2013; Feresntein, 2013). They ran an online experiment to study factors that influences popularity of articles.

First they posted some news stories on a certain online aggregator. An aggregator is a website that aggregates news or other specific content from multiple other web sources. Then they experimentally manipulated the initial number of 'likes' and comments on the news story. They found that articles which had a lot of initial likes and comments tend to have more readability. These articles became more popular as compared to those with low initial likes and comments. Thus an article simply needs fairly good initial popularity and then it is on its way to fame – the 'herd' will do the rest. Something similar may be happening in science – where a couple of initial citations may enhance the article's popularity. And popularity may be mistaken with credibility. This way, herding contributes to the popularity of scientific ideas.

Bear in mind, herding does obstruct scientific progress since it enforces many biases into the dissemination of scientific ideas. Herding is a natural human inclination. Herding could occur due to social pressure. In order not to be an outcast in the community, members often simply mimic the herd. Besides, there is the notion of safety in numbers. Scientists feel the risk is distributed among their large research cluster making them feel safer. This way they become more bold, confident and often increasingly dogmatic (refusing to accept other newer ideas). Avoiding herd mentality is definitely the path of most resistance – it is easier to just slip into the herd. But to promote the progress of science, we would need to minimize things like herding. To overcome herding, scientists would have to do their homework and be more independent in their analyses. They should embrace the risk of being wrong and develop their own scientific convictions without adhering to any 'prominent' schools of thought. When it comes to expert opinion in specific fields, scientists should not rely on sole authorities on the subject but rather seek out second or third opinions.

Along with herding, science is also influenced by 'zombie ideas'. Zombie ideas are concepts which were long considered dead but for some reason get revived. These ideas make a modern comeback poking at current well established scientific concepts.

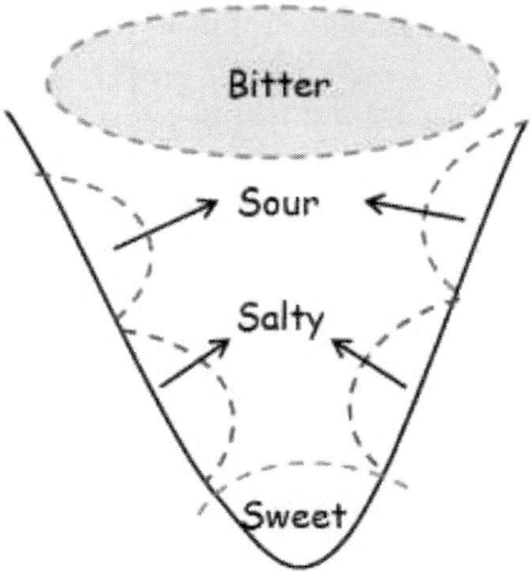

Figure 4.13. Tongue Map.

Among these zombies is the 'Flat Earth Concept', although it has long been proven that the earth is spherical (Dure, 2016). Another is the widely known human tongue map shown in Figure 4.13. The tongue map was first presented in a paper by Ewin G. Broing, a Harvard psychologist – translated from a German article by David Hänig (Hänig et al., 1930). Ambiguously interpreting data from experiments, that paper showed that each part of the tongue tastes only one basic taste. No part of the tongue is localized to a single taste. Though some parts of the tongue may slightly vary in sensitivity to certain tastes, all regions in the tongue could sense all the basic tastes. This erroneous tongue map somehow caught on and went into many biology textbooks. Even today, the tongue map idea is not completely dead – it lingers around like a zombie. Sometimes you may just hear someone bring it up. The same goes with the Flat Earth concept, these zombie ideas can never completely die. Finally there is the ideas linking Measles, Mumps and Rubella (MMR) with vaccines and autism. This is also a zombie that never quiet goes away since its appearance in the prestigious *Lancet* medical journal in 2010. So why do these zombie ideas keep lingering around?

One way to look at the evolution of ideas is to think of it in economic terms (Poole, 2016). If we treat ideas like products (or commodity) which could be bought or sold, then the good products would be traded more; the reverse for bad products. Therefore, ideas which have certain fitness in terms of its usefulness (and perhaps truthfulness) would probably be deemed more fit and thus more likely to be traded and passed on. This way the less fit ideas would be put aside to slowly left to die. In a way the ideas sort of compete with each other to survive. Just like an economic system, the market of ideas sort of self-regulates, filtering-off the rubbish and letting the good ideas live on. In economics, this self-regulation is called the 'efficient market hypothesis'. The efficient market hypothesis is a zombie idea by itself. After experiencing numerous market crashes, it is now known that the efficient market hypothesis is not true. Governance needs to be in place to control and regulate the economic behavior of markets. Not doing this would lead to more financial meltdowns. Like other zombie ideas, the efficient market hypothesis still sometimes makes an appearance. Perhaps the reason zombie ideas continue to exist is because there is a need for them. The efficient market hypothesis is still around because it benefits certain groups; like bankers, huge corporation and people working in the financial sector. To do what they do they would need to believe that the system works; trust and hope that the market would self-regulate and benefit them. A strictly regulated market works against them, restricting their investments and trading activities (Quiggin, 2010).

Similar to economic ideas, perhaps scientific zombie ideas linger because people still benefit from their existence. But how could anyone profit from something like a tongue map? In the book The Tipping Point, Malcolm Gladwell introduced something called the 'stickiness factor' (Gladwell, 2000). The stickiness factor describes how well an idea 'sticks' to our minds and thus gets shared and passed on. The stickiness factor can be increased by making the idea have the following attributes:

1. Attractive
2. Simple; easy to understand
3. Coherent

4. Suggestive
5. Fits well with other ideas; and preconceptions.

Albeit having all those attributes, a sticky idea does not need to be true (adhering to any form of standards). Stickiness might be another reason why zombie ideas exist. This is perhaps why conspiracy theories are so attractive. Zombie ideas often latch themselves on conspiracies. Conspiracies in some ways provide comfort – presenting a grander scheme and providing reasons where none exist. No matter how outlandish and far-fetched the idea may seem, as long as it's sticky enough, a zombie shall rise. No matter the rigor scientists put in publication of science, certain biases and tendencies like herding could not be completely avoided. The same goes with zombie ideas which may be significantly reduced by keeping high standards in scientific facts, but can never be completely eliminated. These tendencies have deep psychological roots and thus will not be so easily circumvented.

Robot Surgeons

Ideas (zombie or not) have never halted the overall progress of the scientific enterprise. When science gets applied, it eventually spills over into engineering. Due to the rapid advances in engineering, many aspects of our modern lives have become automated to a certain degree. With the advent of robotics, the automation revolution has become almost unstoppable, extending its metallic grasp to medical applications – especially surgery. The idea here is to remotely control a robotic system performing surgery with greater accuracy and efficiency compared to direct contact by a human surgeon. Particularly targeted towards complex surgical procedures where the human surgeon is at a disadvantage.

In 1985, the first robot-assisted surgery was conducted. The surgery was minimally invasive as it was performed with very little incisions to prevent extensive scarring and other complications. The system that did the job was the PUMA 560 robotic arm. Five years later, another system,

AESOP was developed – and this time it got (US Food and Drug Administration) FDA approval. Similar to the PUMA 560, AESOP was limited to minimally invasive surgical procedures.

A decade later, in 2000 the FDA-approved da Vinci system hit the market (Lanfranco et al., 2004). The da Vinci system was not just a robotic arm like PUMA 560 and AESOP, it came with the whole works – surgical instruments, camera and optical scopes, all state-of-the-art equipment. da Vinci was a near-complete setup, it did not require much other surgical instruments for support. Besides, it came with a high-resolution camera; this way the surgeon could see the operative region in three-dimensions using powerful magnification capability. Unlike the previous systems, da Vinci's robotic arm is only one centimeter in diameter. This advancement reduces the need for many other surgical instruments. Lesser surgical instruments translate to lesser contact between the instrument and the patient – this significantly reduces the risk of infection during surgery. The da Vinci system was definitely a giant leap in surgical robotics. The operating arm of the da Vinci system closely mimics the motion of the surgeon, making it very easy for the surgeon to control, especially in situations involving tight spaces and require a great deal of accuracy.

Impressive as it may sound, robotic surgery still has a long way to go in terms of completely emulating one of the surgeon's most critical sense – touch (see Figure 4.14). As for vision, the da Vinci system has certainly outdone itself compared to its predecessors with its high-resolution camera and three-dimensional visualization features. Touch is a very complex and rich human sense. The human touch can sense pressure, temperature, vibration, roughness and even tiny shear forces. Surgeons use all these tactile sensations during surgery. The problem with robotic systems is that they do not accurately represent tactile sensations – like the hardness and roughness of surfaces or the organ's minute vibration. Robotic systems tend to over or under compensate these features during surgery. Surgeons thus face difficulty in gauging these features using these robots. Current research shows that our touch perception is so sensitive that it could perceive ridges up to nanometer levels (Skedung et al., 2013). Nanometers are almost the size of individual molecules. Mark Rutland, a Swedish professor specializing in surface chemistry at the KTH Royal Institute of Technology draws an interesting analogy. Consider our fingers to be as big as the earth. It could then still feel individual cars and houses. That's how sensitive our sense of touch is. Our tactile sense comes from organs called mechanoreceptors. These receptors are present beneath our skin, located at different depths from the skin surface. Mechanoreceptors have various sensitivities, some detects the object's shape and roughness while others even sense vibrations.

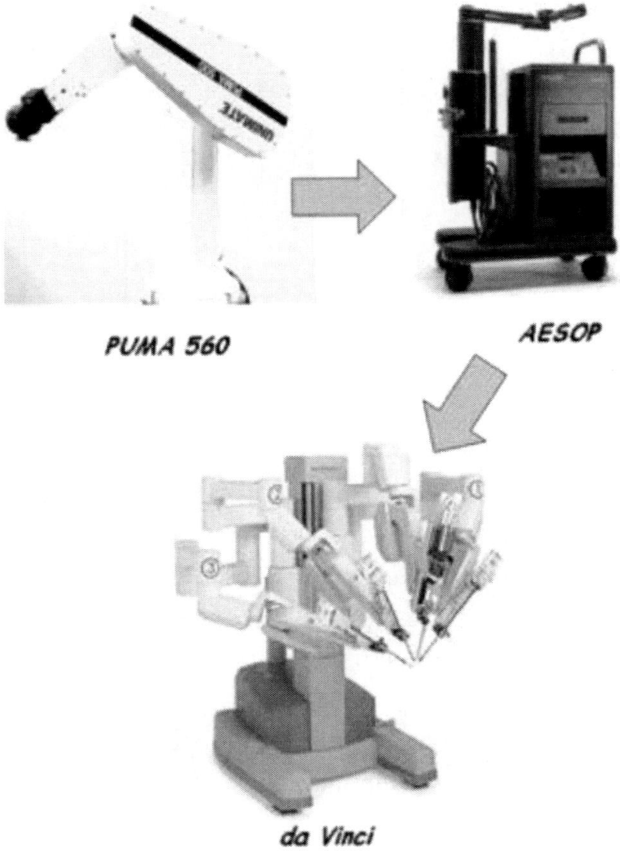

Figure 4.14. Evolution of Surgical Robots.
(Sources: RBO, Technical University of Berlin, 2017; NASA, 2017; Intuitive Surgical, 2017)

This is the type of sensory framework the robotic surgery system aims to have. This way the robot sensors can pick up features of the surface and send it to the surgeon as touch feedback (or haptic feedback). It would be as though the surgeon is doing the operation without the robotic system. Although much improvement needs to be carried out on the robotic system's tactile perception, some physicians have innovated the way surgical robotic systems could be used. One such physician is Dr. Nikolas Blevins (Markoff, 2014). Blevins is a head and neck surgeon at Stanford Health Care. He usually performs very high-precision ear surgeries. During these surgeries, he shaves away bone in the ears so delicately such that the inner surface is as thin as the membrane in an eggshell. Blain is thinking of using robot system like da Vinci for surgery simulations. This way they could try out procedures before doing the real thing. Working with two robotic experts; J. Kenneth Salisbury and Sonny Chan, the group is designing the software for their robotic system enabling it to be used as a simulator. As an input they would have to virtually reconstruct the inner ear in three-dimensions. They did this by using extensive data from inner ear MRI and X-ray images. They hoped that the simulator would be able to mimic different features of the inner ear – such as soft tissue, cartilage and bone. This way surgeons could actually get the real-feel of doing surgery on the simulator. However, the group claims that much improvement is yet to be done on the software to close the gap between simulation and reality.

Researchers in Berkeley took a whole new approach towards the problem. They tried to make the simulator more realistic by increasing its learning capability by letting it tap computational power from the internet – an approach they termed 'cloud robotics' (Kehoe et al., 2015). In another collaborative effort, robotic experts have also developed a database which is connected to the robotic system. This way it could have access to a vast pool of information (videos and images) so that it can operate more smoothly in the real-world.

Robots use computers as brains and unlike humans, computers tend to have a different sort of intelligence. As we have seen, they can freakishly sieve through hundreds of gigabytes of data without breaking a sweat and see patterns where we see nonsense. This is why artificial intelligence experts have much difficulty teaching computers to do simple things most humans reflexively do. Moravec's Paradox described by Hans Moravec in 1990 describes the conundrum of how complex and evolved artificial intelligent systems fail at simple human tasks. As described in his book, 'It is comparatively easy to make computers exhibit adult-level performance on intelligence tests or playing checkers, and difficult or impossible to give them the skills of a one-year old when it comes to perception and mobility' (Moravec, 1990).

Software aside, hardware is also an important factor for surgical robotics. Scientist at Georgia Tech have recently devised sophisticated gadgets for increasing the sensitivity of the robotic sensors (Wu et al., 2013). The team developed bundles of tiny transistors called 'taxels'. The taxes measure mechanical strain and stresses by sensing differences in electrical charges. Using technologies such as this, highly sensitive tactile sensors are integrated into the existing robotic system. The same company that developed the da Vinci system went further to enhance the high resolution visualization system – the upgraded version is called da Vinci Xi.

THERAPEUTIC ROBOTS

The toy company Hasbro recently released their new product, a robotic cat called Joy For All Companion Pets – costing about 99 USD. The cats were tried out at the Hebrew Home, Riverdale, US. The Hebrew Home is a nursing facility for the elderly suffering from any form of dementia (Newman, 2016). People with dementia often have periods of agitation and anxiety. Facilities like the Hebrew Home often employ pet animals or provide privacy (and sometimes isolation) to calm the patients. If the patients go into a full-blown panic, administering tranquilizers are not unheard of. However, often pet animals themselves could get agitated or irritated. This could further aggravate the stress faced by the patients. This sort of situations could be completely avoided with robotic pets. The caregivers at the facility noticed that the presence of Hasbro's robot cats had the same effect on the patients as real pets. Having pets around (real or not) provides the patients with a sense of purpose since it turns them into caregivers. Pets stimulate their senses making them active providing an empowering effect. Many other such facilities have acquired their own robot pets.

Robots are also getting increasingly involved in autism therapy. In the past couple of years, robots have been introduced as part of therapy for children suffering from autism spectrum disorder (ASD). Autistic children are cognitively as well as socially impaired – facing various difficulties in their everyday life. Up until now, there has been a couple of robot therapist models; NAO, FACE (Facial Automation for Conveying Emotions), Roball and Zeno. These robots are designed with sensors, cameras and computers; specially for helping autistic children. The robots help foster social skills among children with ASD. Ordinary children naturally and quickly learn social skills, picking up correct emotional queues and responding accordingly. This is not so with kids having ASD. Social situations often agitate and confuse them. Joshua Diehl, a psychologist at the University of Notre Dame is among the leading experts conducting research on these therapeutic robots. Diehl's team ran an experiment on 19 children with ASD aged between 6 and 13. These children underwent 12 hours of behavioral therapy completing a 1-hour session each time. Half of these sessions were conducted by a human therapist where the other half was done with the presence of a robot. The team even attached biosensors on the children's wrists to gauge their emotional level during the sessions. Diehl and his team found that exposed to the robots, autistic children significantly improved their social skills, overcoming their communication disabilities. He still cautioned that the benefits are individual–specific and cannot be generalized (Diehl et al., 2012; Diehl et al., 2013; Ioannou et al., 2015).

Therapeutic robots would need to be adaptive so that they can deliver personalize therapy to individual autistic children. To approach this issue, a team in Vanderbilt University consisting of autism experts and engineers designed the robotic system ARIA – standing for Adaptive Robot-Mediated Intervention Architecture. The research team headed by Nilanjan

Sarkar, a Professor of Mechanical and Computer Engineering at Vanderbilt University developed this intelligent architecture around the existing robot, NAO (Salisbury, 2013). In the ARIA architecture, NAO stands on a table being eye level with a sitting child. The child is then surrounded by several webcams which monitor movement of the child's head. Using this data, the system can ascertain what the child is looking at. NAO could also produce simple verbal responses. The child is then shown videos during the session. Depending on the child's response the ARIA architecture will intelligently support the child by verbal response (e.g., with phrases like "look over there" or "good job") or using physical gestures. But the level of NAO's interaction is determined based on the intelligent assessment of the child's response using the ARIA architecture. The Vanderbilt researchers also found that the children spent more time looking at the robot therapist as compared to the human therapist. The children were apparently more comfortable with the robot.

In addition to being a co-therapist, robots could also be used as ASD detection tools. Although autism is an incurable life-long disease, recent evidence has shown that its effects could be somewhat minimized with early intervention. To intervene with the development of autistic children at an early age, initial detection is key. This is where the robot comes in. Like many other diseases, early-detection of autism often eludes us. Most of the time, autism is diagnosed after the child is three-years old because conventional autism diagnosis is based on speech exercises. Until the child could speak properly, diagnosis could be extremely difficult if not entirely possible. Recent research has shown that analysis of eye movements in babies could be used to detect autistic tendencies. Such detection techniques use other data like: autism family history and behavior (Cabibihan et al., 2013). Once detected, the therapists as well as robots like NAO could be used to improve their social and cognitive skills – in addition to improving their motor control. Besides eye movement analysis, newer robots like Zeno could detect autism using multiple non-verbal cues such as; body movements and facial expressions (Ranatunga et al., 2013; Tucker, 2015). The robot Zeno was developed by Dan Popa from the University of Texas, Arlington and David Hanson, owner of Hanson

Robots. Besides detection, similar to robots like NAO, Zeno could also act as a therapist. Therapists can program Zeno to behave in a certain way during therapy. They could also control Zeno remotely in real-time by being at a distant from the child. Zeno also has additional feature – it can be controlled by the child. This feature is usually used for entertainment purposes. Inspired by the effectiveness of Zeno the team went on to create another robot, Milo. Outwardly Milo looks similar to Zeno with the exception that Milo could produce facial expressions – emulating emotions. The children's reaction to facial expressions and emotions by robots like Milo is critical data for studies involving disorders besides ASD, e.g., Down's syndrome and Attention Deficit Hyperactivity Disorder (ADHD).

Paralyzed patients have also benefited from robotics. Robotic suits have been developed and tested on paralyzed patients helping them recover some, if not all their lost physical functions. These types of robot suits are exoskeletal structures the patients wear. The structures are connected to electrodes placed on the patient's scalp for detecting electrical signals from the brain. The signals from the brain then tell the exoskeleton the amount of additional support the patient needs to move the paralyzed part of the body. Compensating for the patients loss in strength, the robotic suit helps the patient move. This sort of supportive motion becomes a kind of therapy for fighting paralysis – helping the patient build muscles as well as training the brain and the nervous system to mobilize the body (Donati et al., 2016).

Researchers from Duke University in Durham, North Carolina, started observing patients recovering from paralysis after 12 months of training with robotic suits. Although complete recovery was not achieved, the researchers stated that the improvements at least reduced the severity of the paralysis improving their quality of life. The study focused on patients suffering from paralysis resulting from spinal cord injury. Their findings were consistent with previous scientific understanding that: intensive training could help regain body functions lost from severe injury. This type of training 'rewires' the nerves and brains helping the patient to regain control. This process is termed 'neural plasticity'- the reorganization of the neurons in the brain to perform a certain function (Tublitz, 2008; Kolb and

Whishaw, 1998). The researchers used a virtual reality simulation, before their patients started using the robotic suits. During the simulation, the patient will walk around using their simulated legs. The neuron they use for walking will start firing activating certain parts of the brain. These signals would be learned by the computer connected to the exoskeleton – done using machine learning software. Then, when they actually use the suit, the computer will compensate the exoskeleton accordingly based on the brain signals the computer learned during the simulation.

IN BRIEF

The advent of chaos theory in physics provided new ways of thinking across the life sciences. Organs like the heart and the brain are seen to show chaotic characteristics. These discoveries led to novel early-detection technologies for ailments like Parkinson's disease and migraines. Further influence of the physical sciences is seen in the Miller-Urey experiment – demonstrating the possibility of a completely physical and chemical origin of life on earth. Ideas from chaos theory are closely linked to complex systems like those occurring in the natural world. Self-organization explains natural complexity arising in insects like termites and ants. The notion of attractors help describe the spread of epidemics, as well as how ecosystems move through various equilibriums (e.g., Peter Lake Experiment). Studies on complex systems have spun-off ideas like 'universality' – giving us insight into the way cancer cells metastasize. Thanks to the mathematical framework of Game Theory, the rise and fall of individual animal populations could be understood using the hawk-dove model. Computers on the other hand have become an important addition to the instruments used by biologists and physicians. Computer programs using evolutionary algorithms and machine learning (neural networks) have become indispensable; especially when dealing with large chunks of data. Such programs could run interesting experiments in completely virtual environments. Combing through vast amounts of genetic data, they could also be used to detect certain types of cancers. We see how the physics of electricity has influenced the development of biology from the time of Volta and Galvani right up to recent times; in regeneration research. This chapter ends with some discussion on the social aspects of scientific ideas and a brief overview of cutting-edge robotics in medicine.

REFERENCES

Arnold, C., (2015). Ants Swarm Like Brains Think: A neuroscientist studies ant colonies to understand feedback in the brain, *Nautilus*, no 23: http://nautil.us/issue/23/dominoes/ants-swarm-like-brains-think-rp.

Austad, S.N., 1983. A game theoretical interpretation of male combat in the bowl and doily spider (Frontinella pyramitela). *Animal Behaviour,* 31(1), pp.59-73.

Bennett, K., (2010). *The chaos theory of evolution, New Scientist*: https://www.newscientist.com/article/mg20827821-000-the-chaos-theory-of-evolution/

Berquist, R.M., Galinsky, V.L., Kajiura, S.M. and Frank, L.R., 2015. The coelacanth rostral organ is a unique low-resolution electro-detector that facilitates the feeding strike. *Scientific reports,* 5.

Bittel., J., (2017). *Grow with the flow: How electricity kicks life into shape, New Scientist*: https://www.newscientist.com/article/mg23331050-200-grow-with-the-flow-how-electricity-kicks-life-into-shape/?utm_source =NSNS&utm_medium= ILC&utm_campaign=webpush&cmpid=ILC %257CNSNS%257C2016-GLOBAL-webpush-bioelectricity

Bosl, W., Tierney, A., Tager-Flusberg, H. and Nelson, C., 2011. EEG complexity as a biomarker for autism spectrum disorder risk. *BMC medicine,* 9(1), p.1.

Cabibihan, J.J., Javed, H., Ang, M. and Aljunied, S.M., 2013. Why robots? A survey on the roles and benefits of social robots in the therapy of children with autism. *International journal of social robotics,* 5(4), pp.593-618.

Casey, B., (2016). *Could machine learning diagnose Alzheimer's disease?:* http://www.auntminnie.com/index.aspx?sec=ser&sub=def&pag=dis&ItemID=114592.

Carpenter, S.R., Cole, J.J., Kitchell, J.F. and Pace, M.L., 2010. Trophic cascades in lakes: lessons and prospects. *Trophic Cascades: Predators, Prey and the Changing Dynamics of Nature,* pp.55-69.

Carpenter, S.R., Cole, J.J., Pace, M.L., Batt, R., Brock, W.A., Cline, T., Coloso, J., Hodgson, J.R., Kitchell, J.F., Seekell, D.A. and Smith, L., 2011. Early warnings of regime shifts: a whole-ecosystem experiment. *Science,* 332(6033), pp.1079-1082.

Collij, L.E., Heeman, F., Kuijer, J.P., Ossenkoppele, R., Benedictus, M.R., Möller, C., Verfaillie, S.C., Sanz-Arigita, E.J., van Berckel, B.N., van der Flier, W.M. and Scheltens, P., 2016. Application of machine learning to arterial spin labeling in mild cognitive impairment and Alzheimer disease. *Radiology,* 281(3), pp.865-875.

Coope, G.R., 2004. Several million years of stability among insect species because of, or in spite of, Ice Age climatic instability?. *Philosophical Transactions of the Royal Society of London B: Biological Sciences,* 359(1442), pp.209-214.

Cosarinsky, M.I. and Roces, F., 2012. The construction of turrets for nest ventilation in the grass-cutting ant Atta vollenweideri: import and assembly of building materials. *Journal of insect behavior,* 25(3), pp.222-241.

Crichton, M., (1990). *Jurassic Park,* Barnes & Noble.

Dai, S. and Schaeffer, D.G., 2010. Chaos for cardiac arrhythmias through a one-dimensional modulation equation for alternans. *Chaos: An Interdisciplinary Journal of Nonlinear Science,* 20(2), p.023131.

Davis, M.B., 1983. Quaternary history of deciduous forests of eastern North America and Europe. *Annals of the Missouri Botanical Garden,* pp.550-563.

Dibble, C.J., O'Dea, E.B., Park, A.W. and Drake, J.M., 2016. Waiting time to infectious disease emergence. *Journal of The Royal Society Interface*, 13(123), p.20160540.

Diehl, J.J., Schmitt, L.M., Villano, M. and Crowell, C.R., 2012. The clinical use of robots for individuals with autism spectrum disorders: A critical review. *Research in autism spectrum disorders,* 6(1), pp.249-262.

Diehl, J.J., Crowell, C.R., Villano, M., Wier, K.G., Tang, K., Van Ness, M., Flores, J., Freeman, T., Klinepeter, E.A., Matthews, S. and Mazur, S.L., 2013. The use of humanoid robots as co-therapists in ABA therapy for children with autism spectrum disorder. In *Abstract book for the International Meeting for Autism Research*, San Sebastian, Spain.

Dizikes, P., (2011). *When the Butterfly Effect Took Flight,* MIT Technology Review: https://www.technologyreview.com/s/422809/when-the-butterfly-effect-took-flight/

Donati, A.R., Shokur, S., Morya, E., Campos, D.S., Moioli, R.C., Gitti, C.M., Augusto, P.B., Tripodi, S., Pires, C.G., Pereira, G.A. and Brasil, F.L., 2016. Long-term training with a brain-machine interface-based gait protocol induces partial neurological recovery in paraplegic patients. *Scientific reports*, 6.

Dure, Beau, (2016). Flat-Earthers are back: 'It's almost like the beginning of a new religion', *Science,* The Guardian: https://www.theguardian.com/science/2016/jan/20/flat-earth-believers-youtube-videos-conspiracy- theorists.

England, J.L., 2013. Statistical physics of self-replication. *The Journal of chemical physics,* 139(12), p.121923.

Ferenstein, G., (2013). *The Science Of Reddit: Why Some Ideas Dominate The Net,* The Crunch: https://techcrunch.com/2013/08/11/reddit-science-herd/

Friedman, J., (2014). Science is Not a Popularity Contest, *The Huffington Post*: http://www.huffingtonpost.com/john-friedman/science-is-not-a-popularity-contest_b_5315241.html.

Fritsch, A., Höckel, M., Kiessling, T., Nnetu, K.D., Wetzel, F., Zink, M. and Käs, J.A., 2010. Are biomechanical changes necessary for tumour progression? *Nature Physics,* 6(10), pp.730-732.

Fröhlich, F. and McCormick, D.A., 2010. Endogenous electric fields may guide neocortical network activity. *Neuron*, 67(1), pp.129-143.

Ganesan, T., Vasant, P. and Elamvazuthi, I., (2016). Advances in Metaheuristics: *Applications in Engineering Systems,* CRC Press, Taylor and Francis.

Gaskell, A., 2016. Machine Learning and Hunt for Dementia, *The Huffington Post*: http://www.huffingtonpost.com/adi-gaskell/machine-learning-and-the-_b_12652122.html.

Garcelon, D.K., 1990. Observations of aggressive interactions by Bald Eagles of known age and sex. *The Condor,* 92(2), pp.532-534.

Gardner, M., 1970. Mathematical games: The fantastic combinations of John Conway's new solitaire game "life". *Scientific American,* 223(4), pp.120-123.

Gavrilles, B., (2016). Study explains factors that influence the timing of infectious disease outbreaks, *UGA today*, http://news.uga.edu/releases/article/factors-influence-timing-infectious-disease-outbreaks/

Gherardi, M. and Lagomarsino, M.C., 2015. Characterizing the size and shape of sea ice floes. *Scientific reports*, 5.

Gladwell, Malcolm (2000). *The Tipping Point: How Little Things Can Make a Big Difference*. Little Brown.

Golestani, A. and Gras, R., 2014. Can we predict the unpredictable?. *Scientific reports*, 4.

Gordon, D.M., Dektar, K.N. and Pinter-Wollman, N., 2013. Harvester ant colony variation in foraging activity and response to humidity. *PloS one,* 8(5), p.e63363.

Graber, C., (2016). Replicating Life in Code*, Nova Next*: http://www.pbs.org/wgbh/nova/next/evolution/ai-biology/

Hansen, A.J., 1986. Fighting behavior in bald eagles: a test of game theory. *Ecology,* 67(3), pp.787-797.

Hirota, M., Holmgren, M., Van Nes, E.H. and Scheffer, M., 2011. Global resilience of tropical forest and savanna to critical transitions. *Science,* 334(6053), pp.232-235.

Holland, J.H., 1984. Genetic algorithms and adaptation. In *Adaptive Control of Ill-Defined Systems* (pp. 317-333). Springer US.

Howard, B.C., 2015. How Ants Survive Flooding by Forming Giant Rafts, *National Geographic:* http://news.nationalgeographic.com/2015/10/151006-fire-ants-rafts-south-carolina-flooding/

Hunt, T.M., 1901. *Regeneration*, New York, The Macmillan Company.

Ioannou, A., Kartapanis, I. and Zaphiris, P., 2015, October. Social robots as co-therapists in autism therapy sessions: a single-case study. In *International Conference on Social Robotics* (pp. 255-263). Springer International Publishing.

Ives, C., (2004). 'Human beings as chaotic systems. *Life Science Tehcnology.*

Johnson, A.P., Cleaves, H.J., Dworkin, J.P., Glavin, D.P., Lazcano, A. and Bada, J.L., 2008. The Miller volcanic spark discharge experiment. *Science,* 322(5900), pp.404-404.

Kehoe, B., Patil, S., Abbeel, P. and Goldberg, K., 2015. A survey of research on cloud robotics and automation. *IEEE Transactions on Automation Science and Engineering,* 12(2). pp.398-409.

Kernick, D., 2005. Migraine—new perspectives from chaos theory. *Cephalalgia,* 25(8), pp.561-566.

King, H., Ocko, S. and Mahadevan, L., 2015. Termite mounds harness diurnal temperature oscillations for ventilation. *Proceedings of the National Academy of Sciences*, 112(37), pp.11589-11593.

Kokko, H., López-Sepulcre, A. and Morrell, L.J., 2006. From hawks and doves to self-consistent games of territorial behavior. *The American Naturalist,* 167(6), pp.901-912.

Kokko, H., Griffith, S.C. and Pryke, S.R., 2014. The hawk–dove game in a sexually reproducing species explains a colourful polymorphism of an endangered bird. *Proceedings of the Royal Society of London B: Biological Sciences*, 281(1793), p.20141794.

Kolb, B. and Whishaw, I.Q., 1998. Brain plasticity and behavior. *Annual review of psychology*, 49(1), pp.43-64.

Krueger, J., (1746). *Beschichte der Erde,* [Coat the earth] Lubetvatbischen Buchhandlung, Helmstädt.

Lanfranco, A.R., Castellanos, A.E., Desai, J.P. and Meyers, W.C., 2004. Robotic surgery: a current perspective. *Annals of surgery,* 239(1), pp.14-21.

Ledford, H., (2007). Electric switch could turn on limb regeneration, *Nature*: http://www.nature.com/news/2007/070226/full/news070226-8.html.

Ledford, H., (2015). Bioelectric signals spark brain growth: Voltage changes coax frog cells to build new brain tissue, *Nature*: http://www.nature.com/news/bioelectric-signals-spark-brain-growth-1.17087.

Markoff, J., (2014). Brainy Yes, But Far From Handy: Building a Robot with Human Touch, Science, *The New York Times*: https://www.nytimes.com/2014/09/02/science/robot-touch.html?_r=2.

Mastin, L., (2009). *The Primeval Soup Theory*: http://www. Physicsoftheuniverse. com/topics_life_soup.html.

Mlodinow, L., (2004). *'Some Time with Feynman',* Penguin Press Science.

Monaenkova, D., Gravish, N., Rodriguez, G., Kutner, R., Goodisman, M.A. and Goldman, D.I., 2015. Behavioral and mechanical determinants of collective subsurface nest excavation. *Journal of Experimental Biology,* 218(9), pp.1295-1305.

Moravec, H., (1990). *Mind Children: The Future of Robot and Human Intelligence,* Harvard University Press.

Muchnik, L., Aral, S. and Taylor, S.J., 2013. Social influence bias: A randomized experiment. *Science*, 341(6146), pp.647-651.

Nadal, C. and Majumdar, S.N., 2011. A simple derivation of the Tracy–Widom distribution of the maximal eigenvalue of a Gaussian unitary random matrix. Journal of Statistical Mechanics: *Theory and Experiment,* 2011(04), p.P04001.

Newman, A. (2016). Therapy cats for Dementia Patients, Batteries Included, *New York Times*, Region: https://www.nytimes.com/2016/12/15/nyregion/robotic-therapy-cats-dementia.html?_r=2.

Pai, V.P., Lemire, J.M., Paré, J.F., Lin, G., Chen, Y. and Levin, M., 2015. Endogenous gradients of resting potential instructively pattern embryonic neural tissue via notch signaling and regulation of proliferation. *Journal of Neuroscience,* 35(10), pp.4366-4385.

Pekkala, T., Hall, A., Lötjönen, J., Mattila, J., Soininen, H., Ngandu, T., Laatikainen, T., Kivipelto, M. and Solomona, A., (2017), Development of a Late-Life Dementia Prediction Index with Supervised Machine Learning in the Population-Based CAIDE Study, *Journal of Alzheimer's Disease,* 55(3): 1055–1067. Institute of Clinical Medicine, Neurology, University of Eastern Finland.

Perna, A., Granovskiy, B., Garnier, S., Nicolis, S.C., Labédan, M., Theraulaz, G., Fourcassié, V. and Sumpter, D.J., 2012. Individual rules for trail pattern formation in Argentine ants (Linepithema humile). *PLoS Comput Biol,* 8(7), p.e1002592.

Poole, S., (2016). Why Bad Ideas Refuse to Die, *The Guardian*: https://www.theguardian.com/science/2016/jun/28/why-bad-ideas-refuse-die.

Popkin, G., (2016). Jammed Cells Expose the Physics of Cancer, *Quanta Magazine*: https://www.quantamagazine.org/20160816-researchers-unpack-a-cellular-traffic-jam/

Prabhakar, B., Dektar, K.N. and Gordon, D.M., 2012. The regulation of ant colony foraging activity without spatial information. *PLoS Comput Biol,* 8(8), p.e1002670.

Quiggin, J., (2010). *Zombie Economics: How Dead Ideas Still Walk among Us,* Princeton University Press.

Ranatunga, I., Beltran, M., Torres, N.A., Bugnariu, N., Patterson, R.M., Garver, C. and Popa, D.O., 2013, October. Human-robot upper body gesture imitation analysis for autism spectrum disorders. In *International Conference on Social Robotics* (pp. 218-228). Springer International Publishing.

Reid, C.R., Lutz, M.J., Powell, S., Kao, A.B., Couzin, I.D. and Garnier, S., 2015. Army ants dynamically adjust living bridges in response to a cost–benefit trade-off. *Proceedings of the National Academy of Sciences*, 112(49), pp.15113-15118.

Sabbatini, R.M.E., 1998. *The Discovery of Bioelectricity;* http://www. cerebromente. org.br/n06/historia/bioelectr_i.htm.

Sadati, M., Qazvini, N.T., Krishnan, R., Park, C.Y. and Fredberg, J.J., 2013. Collective migration and cell jamming. *Differentiation,* 86(3), pp.121-125.

Salisbury, D., (2013). *Research News at Vanderbilt*: https://news.vanderbilt.edu/2013/03/23/robot-helps-children-with-autism/

Salleh, A., (2014). Angry birds show too much war is bad, *News in Science*, ABC: http://www.abc.net.au/science/articles/2014/09/10/4084184.htm.

Semedo, D., 2016. Machine Algorithm Ably Predicts Changes in Lung Function in ALS Patients in Clinical Trial, *ALS News Today*: https://alsnewstoday.com/2016/12/09/origent-data-sciences-and-cytokinetics-present-data-on-computer-model-predicting-outcomes-in-als-clinical-trial/

Singer, E., (2014a). The Remarkable Self-Organization of Ants, *Quanta Magazine*: https://www.quantamagazine.org/20140409-the-remarkable-self-organization-of-ants/

Singer, E., (2014b). Machine Intelligence Cracks Genetic Controls, *Quanta Magazine*: https://www.quantamagazine.org/20141218-machine-intelligence-cracks-genetic-controls/

Skedung, L., Arvidsson, M., Chung, J.Y., Stafford, C.M., Berglund, B. and Rutland, M.W., 2013. Feeling small: exploring the tactile perception limits. *Scientific reports*, 3, p.2617.

Solomon, A., (2016). New tool to help predict dementia risk in older people, *Journal of Alzheimer's Disease*: http://j-alz.com/content/new-tool-help-predict-dementia-risk-older-people.

Spano, D., Heck, C., De Antonellis, P., Christofori, G. and Zollo, M., 2012, June. Molecular networks that regulate cancer metastasis. In *Seminars in cancer biology* (Vol. 22, No. 3, pp. 234-249). Academic Press.

Trepat, X., Wasserman, M.R., Angelini, T.E., Millet, E., Weitz, D.A., Butler, J.P. and Fredberg, J.J., 2009. Physical forces during collective cell migration. *Nature physics,* 5(6), pp.426-430.

Tublitz, N., 2008. Neural plasticity: a window into the complexity of the brain. In *Symposium on the Genesis of Syntactic Complexity,* Rice University, Houston (March 2008).

Tucker, E., (2015). How robots are helping children with autism, Health and Wellbeing, *The Guardian*: https://www.theguardian.com/lifeandstyle/2015/feb/01/how-robots-helping-children-with-autism.

Turner, J.S. and Soar, R.C., 2008, May. Beyond biomimicry: What termites can tell us about realizing the living building. In *First International Conference on Industrialized,* Intelligent Construction at Loughborough University.

Tuthill, K., 2003. John Snow and the Broad Street Pump: On the Trail of an Epidemic, *Cricket Magazine* 31(3), pp. 23-31. Available online at: http://www.ph.ucla.edu/epi/snow/snowcricketarticle.html.

Wells, D.A., 1872. *The science of common things: a familiar explanation of the first,* Ivison, Blakeman, Taylor & Co., 323 pages (page 290).

Wiley-Blackwell. "New research rejects 80-year theory of 'primordial soup' as the origin of life." *ScienceDaily,* 3 February 2010. <www.sciencedaily. com/releases/2010/02/100202101245.htm>.

Wolchover, N., (2014). A New Physics Theory of Life, Physics, *Quanta Magazine*: https://www.quantamagazine.org/20140122-a-new-physics-theory-of-life/

Wolchover, N., (2015). Listening to Nature's Early Warning System May Save Species, *Science*, Wired: https://www.wired.com/2015/11/listening-to-natures-early-warning-system-may-save-species/

Wu, W., Wen, X. and Wang, Z.L., 2013. Taxel-addressable matrix of vertical-nanowire piezotronic transistors for active and adaptive tactile imaging. *Science,* 340(6135), pp.952-957.

Yulmetyev, R.M., Demin, S.A. and Hänggi, P., 2006. Manifestation of chaos in real complex systems: Case of parkinson's disease. In *The Logistic Map and the Route to Chaos* (pp. 175-196). Springer Berlin Heidelberg.

Figures Source

da Vinci System, (accessed in 2017). *Intuitive Surgical*: https://www.intuitivesurgical.com/company/media/images/davinci_si_images.html.

Robotics and Biology Group, (accessed in 2017). Technical University of Berlin: http://www.robotics.tu-berlin.de/menue/home/

Robotic Surgery, (accessed in 2017). NASA Spinoff: https://spinoff.nasa.gov/spinoff2000/hm1.htm.

Chapter 5

STRANGER THINGS IN LIFE

"Seek not the paths of the ancients;
Seek that which the ancients sought."

— Matsuo Basho,
Haiku from Words by a Brushwood Gate

THE FIRST QUANTUM BIOLOGISTS

Three physicists blazed through the first paths leading towards quantum biology. Working in the 1930s, they daringly attempted to understand biological phenomena using a quantum physics viewpoint. The three pioneers were: the illustrious atomic physicist, Niels Bohr, the Nazi scientist, Pascual Jordan and the biophysicist, Max Delbrück. These were the men whose work converged around quantum biology albeit their researches were aimed towards completely different targets. Before we go into the details on how these individuals got involved in quantum biology, let's first take a look briefly at some of the quantum discoveries that spilled over into biology. Among the key ideas proposed by the new quantum physics is the 'complementary principle'. The complementary principle originates from a quantum physics experiment known as the double-slit experiment. The double-slit experiment was first devised in 1801 by

Thomas Young – an English physician. Although being a physician, Young also dabbled in the physical disciplines such as, solid mechanics, energy and research into the nature of light. Young designed the experiment to conclusively demonstrate the wave aspects of light. He set out to show that similar to liquid water, light behaved like a wave when obstructed. For instance, consider water stored in a tank equipped with a wave generator, as shown in Figure 5.1. The wave generator is simply a ball-and-stick device which bobs a ball in and out of the water at fixed periods. This 'bobbing' generates waves, rippling the water as it moves towards the end of the tank. When the waves are obstructed with a wall with two slits (double-slit), the slits act like a new source. Passing through the slits there are now two wave sources. The waves from these sources interfere with each other causing interference patterns. Interference effects provide definitive proof that whatever it is passing through the double-slit is a wave.

Young ran the double-slit experiment using light – done by simply replacing the ball-and-stick mechanism with a light source. Similarly, the light is passed through two slits. Unlike water interference patterns of light cannot be observed from the top. So Young placed a white screen at the end of the setup after the slits (similar to screens in front of a projector).

The interference patterns of light could then be observed on the white screen – seen as black and bright stripes. If whatever being projected is not a wave, then there will be no stripes. Using the double-slit experiment, Young proved that light is indeed a wave. But all this was long before quantum theory became known, which is beginning of the twentieth century. The double-slit experiment then makes a comeback in quantum theory and used to describe the complementary principle. The complementary principle is also known as wave-particle duality because quantum particles could behave like a wave or particle depending on the observer. In such quantum experiments the observer cannot be isolated; and so we enter the realm of science where the objectivity of the observer is in question. When we say quantum particles, we are talking about any material or particles that are extremely small with an approximate scale of below 10^{-9} meters.

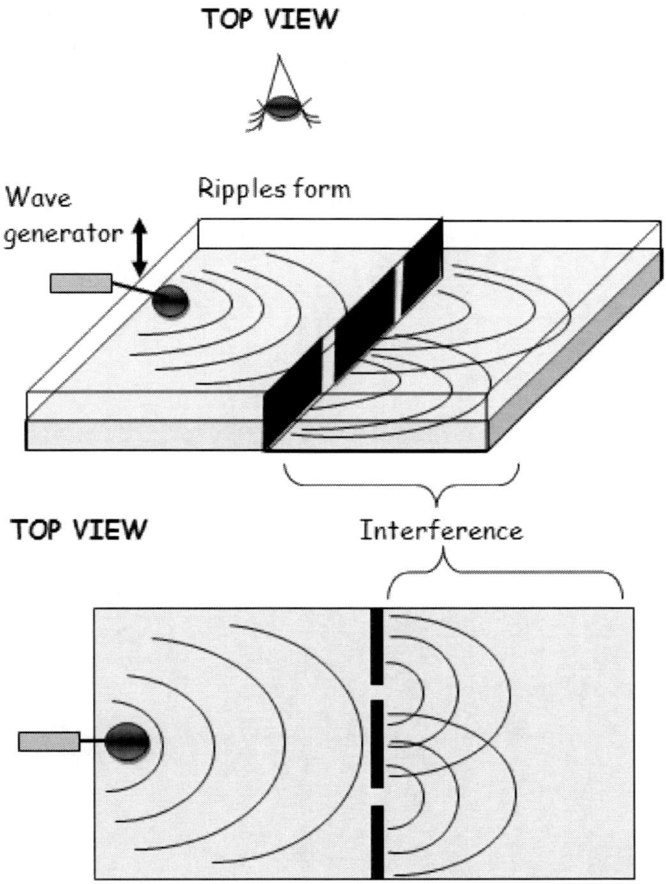

Figure 5.1. Wave generation and interference in liquid water.

In the quantum version of the double-slit experiment, the light source (or ball-and-stick device) is replaced by a new source emitting a single quantum particle (for simplicity let's say this source emits a single electron). Just like the experiment with light, a screen could be placed at the end to see if there is any interference pattern – to show it's a wave. Upon emission, the electron passes through the two slits, as shown in Figure 5.2. The strange thing is that if someone is there observing (or measuring) the electron, it behaves like a particle. On the other hand, if there is no observer present, then the electron moves as a wave. In short, the complementary behavior of the electron is very subjective to the experimenter. The electron behaves like a wave or a particle depending on the way the experimenter asks the question and the chosen method of experiment.

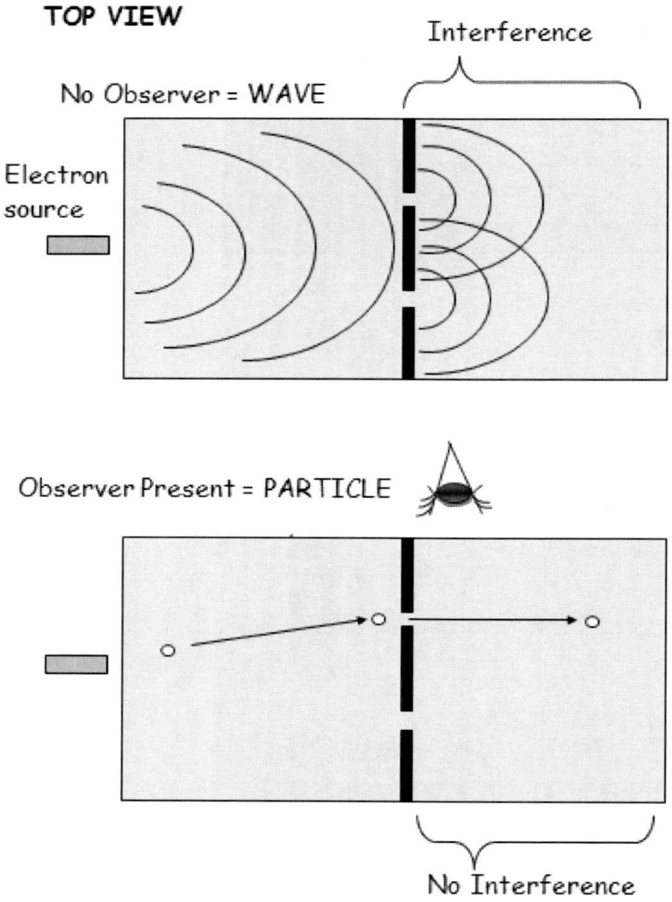

Figure 5.2. Quantum double-slit experiment.

This paradoxical behavior created a stir in the physics community – intriguing many prominent physicists at the time including Albert Einstein and Neils Bohr. Up to that point, science was an objective endeavor. Scientists remained cut-off from the phenomena under study, enabling them to draw conclusions and analyze data objectively – without influencing the outcome of their experiments. As the quantum double-slit experiment came to light, all these ideas got thrown out the window. A new type of subjective science was emerging and the foundation of it was built on the complementary principle. Trying to make sense of all this, Erwin Schrödinger dreamed up the famous thought experiment, called 'Schrödinger's Cat'. Even today, no matter how many times we run the quantum double-slit experiment (using various state-of-the-art technologies for measurement), the complementary principle always holds. At quantum levels wave-particle duality is an indisputable reality – though the exact reason(s) for this remains unclear.

So today scientists continue doing physics or any other related science keeping the complementary principle as a de facto concept – just as in biology, where it's taken that subjects under study have life. These sorts of analogues between quantum physics and biology attracted the interest of Niels Bohr. Although having a keen interest in biology, Bohr never engaged in biological research directly. He explored biology at a philosophical angle via discussions and lectures.

Niels Bohr was the son of the eminent Danish physiologist and professor at the University of Copenhagen; Christian Bohr. Being one of founders of quantum physics, he was famous for developing the atomic model of the hydrogen atom. For his efforts, Bohr was awarded the 1922 Nobel Prize in Physics. As a contemporary of Einstein, they often debated issues in physics – especially quantum theory.

Even as an adolescent, Bohr was exposed to deep discussions covering wide areas in the arts, science and philosophy. These discourses were held at home between his father and his friends. During one of these discussions, Bohr for the first time came to understand the philosophical problems encountered in biology (Bohr, 2017). When he later encountered the quantum double-slit experiment, he saw the complementary principle as a way to understand phenomena – not only in physics, but across other disciplines like biology and psychology. The quantum double-slit experiment tells us that the only way to understand electron movement (or any other quantum particles) is to accept the fact that electrons have two forms; wave and particle. Using a single perspective we can never completely comprehend what's really happening. Bohr was deeply inspired by this mode of thinking.

Applying it to biology opens up two ways of understanding life. Consider the hunting of gazelles by cheetahs. This scenario could be understood at a *molecular level*; where the cheetah rapidly increases its metabolism causing high-speed chemical reactions and electrical signals to surge through its body pushing it to achieve extreme velocities for capturing the gazelle. Another view is to understand the *climate and resources* influencing the availability of prey using *population statistics* of cheetahs and gazelles in the region. Bohr's reasoning could also be extended to social psychology. The first approach to understand the behavior of a person is to observe genetic and environmental factors affecting the individual. A complementary method is to analyze population-scale trends like influences of culture, religion and other social statistics. The philosophy of the complementary principle was presented by Bohr in a lecture called 'Light and Life'. This opening address was delivered in 1932 at a meeting called International Congress of Light Therapy in Copenhagen. In the lecture Bohr proposed that scientists should focus on problems where the results from physics could be applied to understand living organisms (Joaquim et al., 2015).

Bohr invited his longtime friend and 'disciple', the then budding physicist Max Delbrück. Inspired by Bohr, he will later become the most committed quantum biologist of his time. Delbrück was born in Berlin in 1906. He was the son of Hans Delbrück, a history Professor at the University of Berlin. Initially studying astrophysics, Delbrück went on to complete a doctorate in theoretical physics. Even in his physics days, Delbrück was interested in the science of genes. He was often occupied in developing the so called 'Delbrück model' – attempting to explain the nature of genes using the newfound quantum theory and its description of matter. Erwin Schrödinger was extremely convinced with the Delbrück model – thinking that the model was our best shot in consolidating quantum physics and biology. He even mentions this in his book – "If the

Delbrück picture should fail, we would have to give up further attempts" (Schrödinger, 1944).

Doing research in Switzerland and Denmark, he collaborated with the leading theoretical physicist, Wolfgang Pauli. This was when he first met Niels Bohr, who will remain an influential figure throughout his life – to the extent that Bohr will invite him personally to attend and become the principal speaker at his 'Light and Life' lecture. To ensure that Delbrück attended the lecture, Bohr sent his collaborator Leon Rosenfeld to pick him up at the Copenhagen train station. Throughout his life, Delbrück was known to declare that Bohr's lecture was his starting point of his career in biological research.

During the Nazi occupation of Germany, Delbrück applied for the Rockefeller fellowship (an American research grant). Upon securing the funds, he left Germany for the States – specifically to California Institute of Technology (Caltech). There he worked on the genetics of fruit flies. He quickly learned that this line of research could never help him fulfill his dream to unite quantum physics and biology; which is to complete the Delbrück model (Fischer, 2007). So he shifts his research angle again, this time studying bacteriophages. Bacteriophages (sometimes simply known as 'phages') are *viruses* that only attack bacteria. This is when Delbrück invokes the complementary principle. He begins considering viruses in genetics as equivalent to atoms in quantum physics. Unlike Bohr, Delbrück did not take a theoretical approach in biology research – he pursued rigorous hands-on experimental work. His work on phage research elevated him to a prominent leader in molecular biology. His enormous contribution to the field won him the 1969 Nobel Prize in Physiology. Delbrück stayed on in the U.S. for the rest of his life. Being an influential physicist-turned-biologist, he encouraged scientific explorations involving the use of physics to deepen our understanding of biology.

Finally we've arrived at the most controversial quantum biologist; Pascual Jordan. Jordan was not as well-known as Bohr and Delbrück. Nevertheless, he was among the pioneers that laid down the foundations of quantum physics. Born in Germany, Jordan studied a mix of zoology, mathematics and physics at the Hanover Technical University. He later switched to Göttingen University to complete his studies focusing on the physical sciences. There he worked with illustrious scientists like Werner Heisenberg and Max Born (who was incidentally the grandfather of the singer and actress, Olivia Newton John). The three men collaborated and produced ground-breaking work in quantum mechanics (Born et al., 1926). Then war burst out throughout Europe and at the center of it all was Nazi Germany. Jordan joined the Nazi party and continued working on science. He knew about Bohr's ideas on how quantum principles could be carried over to biology, as well as other fields. He even had plans to setup a quantum biology institute after the supposed victory of Germany. The institute would also incorporate race research into its study program. Although Jordan was influenced by Bohr's ideas, Bohr himself avoided being connected in any way with Jordan or his views.

All that aside, Jordan was a man of many contradictions. His character was riddled with many conflicting behaviors. On one hand, he was into race research and had the 'Nazi scientist' thing going on. On the other he supported Jewish physicist and scientists like the psychologist, Sigmund Freud. He tried to propose a place for the newfound quantum mechanics in the Third Reich. This idea faced outright rejection since it built on the foundations laid by many Jewish physicists. Besides quantum biology, Jordan had a knack for following unconventional research trends. For example he was interested in movements like organicism. Inspired by Bohr, he tried to introduce the complementary principle into biology.

His approach was called 'Amplifier Theory'. The theory aimed to describe how living systems (like cells and organisms) amplify quantum events occurring within them. According to amplifier theory, living systems bridge the gap between the very small scales (quantum level) and events that occur at larger scales (normal or macroscopic level). The central idea of the theory is to effectively divide a living system (let's say a cell) into two zones. The first zone, operating on a quantum level contains the essence of life (life zone). The second zone receives the amplified commands from the life zone and converts it into cellular functions (metabolism, digestion,., etc.). The second zone functions on a macroscopic level. Jordan's model was speculative at best, although he was hopeful that advancements in physics could revolutionize biology. His amplifier theory shared similarities with his political inclinations. The amplifier theory mimicked the Nazi concept of the Führer; where the power and control of the Führer extends all over Germany and further. The complementary principle as seen and developed by Bohr, Delbrück and Jordan is given in Figure 5.3:

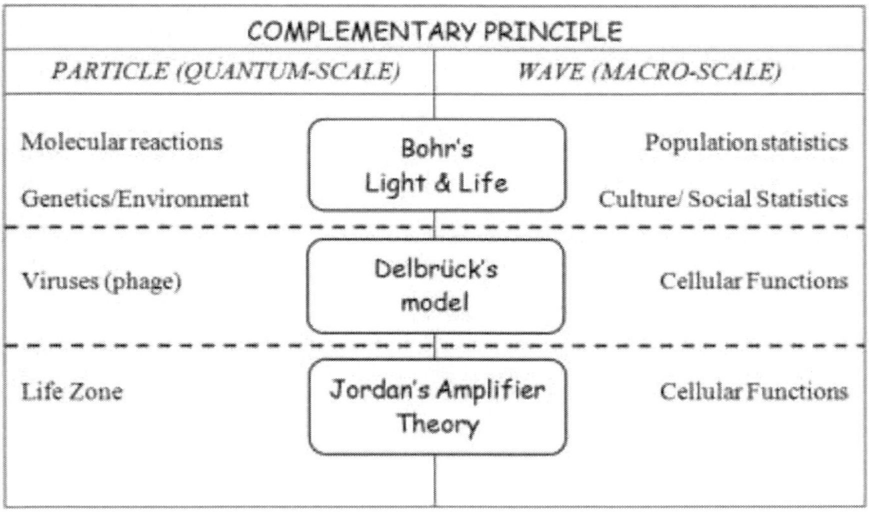

Figure 5.3. The Complementary Principle Represented in Biology.

TECHNOLOGICAL REVOLUTION

Bohr, Delbrück and Jordan never made direct contributions to quantum biology. Their explorations influenced and inspired many scientists to look at biology in the light of new developments in physics. Being the most rigorous of the three, Delbrück certainly made the most advancement, steering his entire scientific career in the direction of biology. This interdisciplinary line of research catalyzed the technologies used in biology experiments. An example is the electrophoresis method used to efficiently separate molecules. This method uses electricity (specifically an electric field) to separate molecules based on size, density and purity. It is especially useful for separating various proteins (Lyklema, 1995; Hunter, 1989) Using this technology, biologists could understand how proteins bond with DNA. Electrophoresis has also been used extensively in research involving antibiotic resistance (Sautrey et al., 2015).

Spectroscopy was developed based on purely physical principles. Before we delve deeper into its application, let's first briefly review how spectroscopy works. Spectroscopy started when scientists studied how visible light got scattered according to its wavelength – like what happens when light passes a prism. As with light scattered by a prism, light used in spectroscopy behaves like a wave – although at quantum levels it may act like a particle known as photons. So the light wave coming out of a prism is separated into its individual colors. Each color is represented by a signature wavelength. This is because white light contains all wavelengths and the prism separates white light into its individual wavelengths. Water droplets also act like prisms and produce rainbows.

Scientists later understood that this same scattering mechanism is what gives objects color. Dyes and paints coming from a material scatter light into a specific wavelength giving it a certain color. Consider a green apple. When exposed to white light, the apple absorbs all other wavelengths and reflects only a single wavelength – which our brain then processes as the color green. The same goes for other wavelengths within our visible range – each wavelength is perceived as a particular color. Light is an electromagnetic wave. When its wavelength becomes too small or too large, it slips out of our visual range. The largest wavelength we can perceive is red while the smallest is violet. After that the electromagnetic wave falls into the infrared (IR) or ultraviolet (UV) range. As the electromagnetic wave reduces in wave length, it becomes microwaves, X-rays and gamma-rays; all of them cannot be perceived visually by us.

Best part is, thanks to spectroscopy we even know why the sky is blue. A phenomenon called Rayleigh scattering – where water, dust and air molecules in the atmosphere scatter light such that only light with the blue wavelength is reflected. Naturally during sunset and sunrise, when the angle of the sun varies, different colors are absorbed and reflected giving us breathtaking views of the sky. Keep in mind that different materials are composed by different chemicals. And these chemicals have specific atomic structures. To understand the way light gets scattered by different materials, we need quantum physics. So by exposing some material to certain forms of electromagnetic waves and studying the properties of scattered waves, we can know the material's chemical type or other characteristics (Figure 5.4).

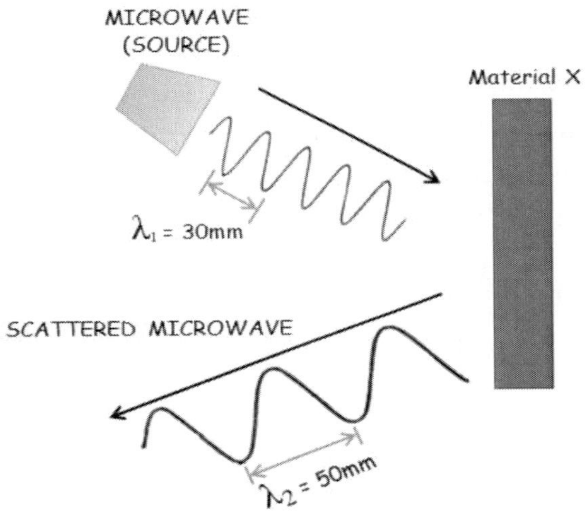

Figure 5.4. Physics of Spectroscopy.

Consider identifying an unknown material called Material X (shown in Figure 5.4). A source microwave radiation is targeted at the material with say wavelength, λ_1= 30 mm. Material X than scatters the microwave resulting in a wave with λ_2= 50 mm. Different types of materials scatter microwave differently causing the scattered wave to have different properties (wavelength, frequency,, etc.). By measuring the amount of increase in the wavelength after scattering we could identify and characterize Material X. This is what spectroscopy is all about and why it remains a very useful tool for characterizing materials. Spectroscopy has been used broadly across various experimental fields. In molecular biology, spectroscopy is often used to determine DNA or RNA concentrations in mixtures. Spectroscopic methods have now become standard practice in molecular biology, catalyzing the pace of experimental research.

The double-helical structure of the DNA couldn't have been discovered without using crystallography. The concept of crystallography is in a way similar to spectroscopy – where it involves bombarding materials with a wave and then analyzing the resulting wave. The difference though is that crystallography is used specifically to determine the molecular or atomic structure of crystals. When we say crystals, what we really mean is solid materials which have a crystalline structure. Examples of crystals include table salt, snowflakes, diamonds and glass. These materials have a regular or periodic molecular structure repeating in the molecule forming a geometric pattern. Figure 5.5 shows the crystal structure of table salt, chemically known as sodium chloride (with the formula NaCl). The sodium and chlorine ions are geometrically arranged repeating its configuration throughout the molecule. Note that we are using the term 'ions' and not 'atoms'. This is because table salt's crystal structure contains ions. Ions are atoms which have gained or lost electrons, rendering them electrically charged. The chemical symbols of sodium and chlorine in Figure 5.5 are marked with negative (-) and positive (+) signs to indicate them as ions. These ions share electrons among each other to form bonds in the crystal.

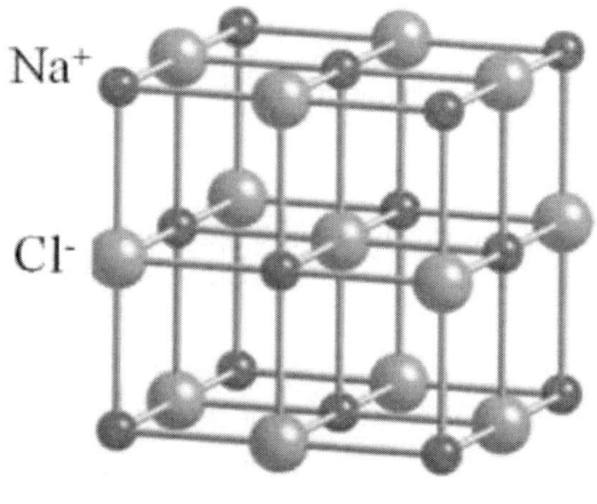

Figure 5.5. Crystal Structure of table salt.

This kind of repeating geometric pattern seen in table salt is also present in all crystals. Unlike spectroscopy, crystallography does not involve wave scattering, it relies on wave diffraction. Wave diffraction is what happens when a wave passes through a slot similar to the double slit experiment like in Figure 5.1, but with only one slit. The wave then interferes only with itself forcing its way through a single slit forming diffraction patterns. So when it comes to crystallography, the wave source should have a wavelength small enough for it to squeeze itself in the spacing between the atoms of a crystal. This is why scientists often use X-rays for crystallography, since they have tiny wavelengths.. By analyzing the diffraction pattern's intensity and angle of diffraction once it passes through a material, the crystallographer can know the density of electrons in the material. With this understanding one can deduce the crystal structure and the atomic configuration of the material.

In Chapter 3, we saw how Schrödinger postulated that genetic material would have a periodic crystalline structure, long before the discovery of the DNA. Betting on Schrödinger, biologists and chemists started picking up crystallography to search for this crystalline structure in biological molecules. Among them was Rosalind Franklin, an English chemist working at Kings College, London. Using X-ray crystallography, Franklin attained images of the DNA. These images were later used in 1953 by Francis Crick and James Watson of Cambridge University to build their atomic configuration of the DNA. Today we know the DNA as a double-helical structure containing over three billion base pairs, all the information required for the functioning of a human being. The human DNA stretches over six feet and is compressed to fit in the nucleus of a single human cell.

The quantum biologists stirred the pot mixing physics and biology – improving each field using ideas from the other. The ripples of what they started then are still felt today. And it is most felt in the past ten years, as scientists begin to literally connect quantum theory and biology. We are on the brink of a quantum age in biology where quantum theory will help us navigate through the complicated world of the life sciences.

QUANTUM EFFICIENCY

Photosynthesis is the first touchdown of energy on earth from the sun. Plants on earth absorb about 10^{17} Joules of energy every second – that is about the total energy produced by 20 million high-capacity nuclear plants per second. And all that energy is taken up by plants which is then passed to animals that eat them, forming earth's energy cycle. But before that, plants have to transform sunlight and carbon dioxide into carbohydrates (life-supporting sugars). What is astonishing though is that plants can do it at as much as 95 percent efficiency. Which means 95 percent of all that sunlight gets converted into carbohydrates. The remaining 5 percent becomes waste energy. Such efficiencies have never been achieved by any man-made device. Even nuclear plants have a maximum conversion efficiency of 37 percent – they only convert 37 percent of nuclear energy to usable electrical energy. The remaining energy is lost as nuclear waste, heat and radiation.

So how do plants do it? Working at University of California, Berkeley, Graham Fleming worked on understanding photosynthesis for more than a decade. Using methods like high-tech laser spectroscopy, the research team tried to track down the energy movement during photosynthesis (Engel et al., 2007). Besides plants, certain types of bacteria (e.g., green sulfur bacteria) also could photosynthesize carbohydrates using light. Instead of plants the team used these bacteria for tests. In 2007, the team zoomed-in and managed to trace the path of a particle of light (photon) as soon as it touches the bacteria. It was expected that the photon gets converted into an electrical charge and then transported to wherever it was needs to go. Similar to how electrical charges flow through a wire. But this was not the case. They observed that no such conversion happened; the photons remained as it were. If the photons are moved like this, having nothing to drive it to its destination, it would end up anywhere randomly. Their experimental results were showing that the photons were defying all odds – almost all the photons were reaching their destination. Further analysis using laser spectroscopy showed that the photons were transported with the help of quantum physics. The electrical charge movement hypothesis was completely thrown off. During its movement the photons were undergoing quantum superposition. Let's explore how quantum superposition works. Previously we talked about the complementary principle; wave-particle duality. Looking back at Figure 5.2, we can only spot the electron or photon once we observe it. Before that it remains wave-like; which means that it could be anywhere in the double-slit setup. Until the measurement is made, there is a chance (probability) for the particle to be somewhere in the setup. Once the measurement is made and the particle is observed, the wave 'collapses' into a particle in the most energy-efficient way. Consider another example depicted in Figure 5.6.

Assuming in this case there is no help from quantum superposition, then the photon moving through the bacteria would have to randomly try out each path: *A*, *B* and *C*. Then choose whichever path that gets it to its destination. Unfortunately if the photons take this path, its energy would easily dissipate along the way causing very few photons to make it to their targeted destination, *X*. With quantum superposition, the photons could take all three paths simultaneously, wave-like. Upon measurement the photon simply collapses to the most efficient path. This way there will be minimal losses of photons and all the energy would be transported optimally to its destination. This is what the Berkeley team discovered with their experiments on green sulfur bacteria. Quantum superposition was boosting up the energy efficiency of plants and bacteria alike during photosynthesis. Besides bacteria, quantum superposition has also been observed in plants like spinach (Schlau-Cohen et al., 2012; McFadden, 2017).

The Berkeley team ran their photosynthesis experiments at extremely cold temperatures. Such temperatures allow quantum superposition to remain stable for an extended time during photosynthesis. A long-lived quantum superposition would thus give the team sufficient time to perform measurements and collect data. The downside is that the data they are getting might not depict the process accurately as it happens in nature. This is because plants don't need cold temperatures to photosynthesize. They easily thrive in a wide range of temperatures. Building on the findings of the Berkeley team, another research group at Toronto University started running their own photosynthesis tests. The research team consisted of the physical chemist, Elisabetta Collini and her colleagues (Cartlidge, E.,

2010; Collini et al., 2010). Instead of using green sulfur bacteria, Collini's team used a type of photosynthesizing algae. In their experiments they managed to show that quantum superposition of the photon paths still happen at ambient temperatures – further strengthening the idea that photosynthesis is a quantum–driven process.

As with plants, animals too have their own quantum tricks for optimizing their energy use. Cells take up energy for their everyday operations. The part of cells which generate all this energy is the mitochondrion. This miniature power plant converts sugars to chemical fuel; adenosine triphosphate (ATP for short). ATP is the cellular version of engine fuels like petrol. ATP is then moved around the cell to be converted into energy to be used for cellular activities like repair or multiplication. So we have two processes required for generating energy for the cell. First being the conversion of sugars to ATP by the mitochondrion. Secondly the transport of ATP to wherever it's needed – to be changed to usable energy. Now all these steps should take a lot of time. This doesn't make sense considering the high speeds at which energy is generated in animal cells; especially the rate at which ATP is churned out. Biologists always thought that electrons 'hop' from one molecule to the next in the mitochondria when making ATP.

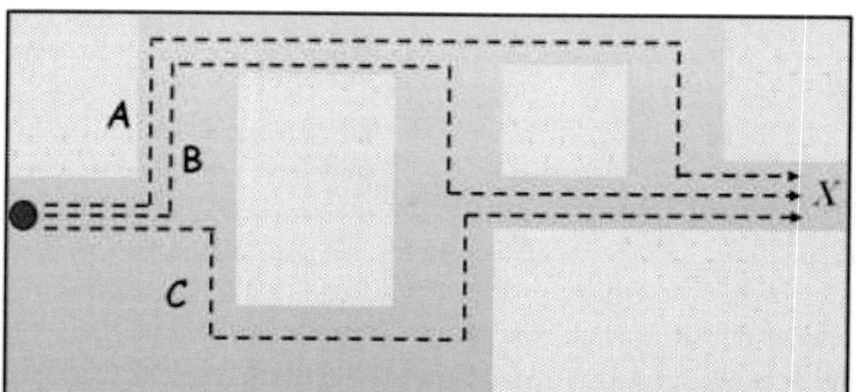

Figure 5.6. Photon Path during photosynthesis.

Inspired by the Berkeley team, Vlatko Vedral, a physicist at the University of Oxford, U.K. challenged this notion (Arndt et al., 2009; Brooks, 2011). By this time, the quantum plague was already starting to infect the research ideas of many scientists. Pursuing this line of thought, Vedral and his team developed a physics model that predicts the mitochondrion's ATP production rate. But instead of using electron hopping, their model used quantum superposition. Similar to the photosynthesis explanation, the superposition approach allows the electrons in the mitochondrion to try multiple routes at once. This way they avoid random testing routes which are energy-consuming and slow. Vedral's physics model was shown to mimic experimental results confirming the quantum explanation for energy generation in animal cells.

EAU DE QUANTUM?

The sense of smell is commonly shared by most animal species across the planet. In the animal kingdom, smell is used for: mating, migration and hunting among many other things. A normal human being can distinguish about a trillion different smells. That's one with twelve zeroes after it, 10^{12}; how about that. Salmon use their sense of smell to navigate upstream against the flow of currents from rapids to reach their spawning ground. More astonishing is the keen sense of the Auburn tracking dog. The dog can follow an individual human trail 24 hours after in a campus frequented by tens of thousands of students (Matthews, 2013). An interesting yet slightly controversial recent development is the quantum explanation of smell called quantum vibration scent theory.

Let's first look at the quantum part of the theory – quantum tunneling. Quantum tunneling is the ability of an electron to simply pass through walls (or barriers of any kind). Looking back at the complementary principle: electrons exist as waves and particles simultaneously. Consider that in Figure 5.2, we use a wall replacing the double-slit. Then prior to measurement (when the electron is a wave), it can be anywhere before or after the wall. The existence of the wall does not hinder the wave from being at any location in the setup. That's quantum tunneling – adding more weirdness to the already weird quantum theory. Now let's return to the quantum scent theory. The theory was first proposed in 1996 by biophysicist, Luca Turin. Turin suggested that quantum theory may be playing a large role in the way we smell. Before Turin, the prevailing explanation was chemistry-based using a lock-and-key approach. This explanation assumes that our nasal system has a built-in library of receptors. The many receptors cater for different smells classified according to the odor molecule's shape and size. When an odor molecule (key) approaches the right receptor (lock), everything clicks. The receptor then sends a signal to the brain and the person gets the smell. Until recently, the lock-and-key approach seemed adequate. However, it starts cracking when dealing with molecules that have different shapes but smell the same. One good example is a molecule, hydrogen sulfide. Hydrogen sulfide comes in a variety of shapes but always smells the same – like rotten eggs. Such situations cast doubt on the lock-and-key model.

According to Turin's quantum scent theory, the shape and size of odor molecules do not affect its smell. His idea is based on molecular vibrations. All molecules have three modes of movement: translation, rotation and vibration. This means they can move following a path in any direction (translate), rotate as well as vibrate according to some frequency. The quantum scent theory is only concerned with the vibration mode. Using the theory, even molecules with various shapes and sizes could be detected by the same smell. This is how Turin's theory works. In the nose there are receptors which look like tuning forks shown in Figure 5.7.

The space between the receptor edges act as a barrier. On one edge of the receptor sits an electron – being in a wave form, it's everywhere at once. When an odor molecule with the right vibration comes along, the electron collapses to a particle; tunneling through to the other edge of the receptor. The vibration of odor molecules behave like an observer measuring the electron. The collapsed electron then sends a signal to the brain registering a smell. Turin's theory is not about shapes and sizes, it's all about the right vibes!

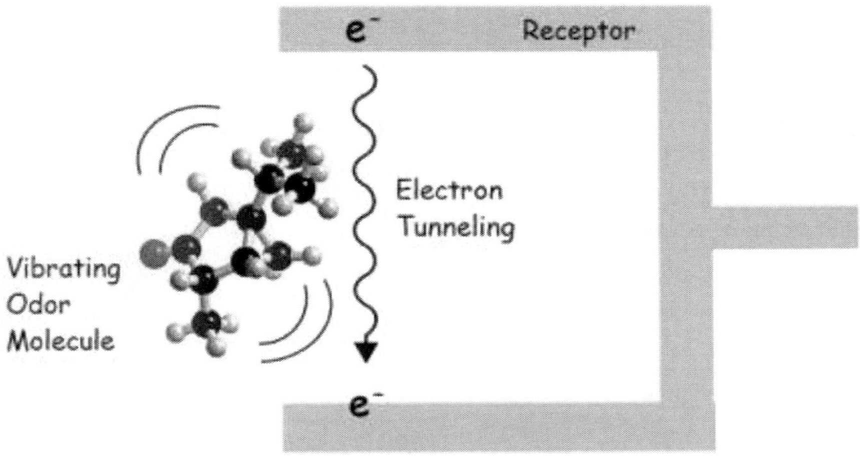

Figure 5.7. Quantum tunneling during scent.

Be that as it may, without conclusive experiments the theory faces many challenges and critics. So Turin and his team designed an experiment using fruit flies since previous research using humans produced inconclusive proof of the theory. The team used two molecules which had the same shape but different vibrations. The funny thing about fruit flies is that they can be trained to recognize different odors. Fruit flies in this experiment were trained to recognize the odor of a particular chemical (let's call this Chemical A). Chemical B has the same shape as Chemical A but vibrates differently. If the quantum scent theory is right, when the fruit fly is exposed to the odor of Chemical B, it would be able to distinguish it from Chemical A. Otherwise it would remain indifferent to both chemicals. The results were that the fruit flies could differentiate the two different odors, proving the validity of the quantum scent theory. Turin later stated – "My theory was described as impossible physically, implausible biologically, not supported by evidence. This is a clear indication that some component of fruit fly olfaction is sensing vibrations" (Courtland, 2011; Franco et al., 2011).

Avian Navigators

Many species of birds migrate every year crossing thousands of kilometers around the globe. Birds usually migrate in search of food or to avoid bad weather. Humans too migrate seasonally to avoid cold weather – often referred to as 'snowbirds'. Birds on the other hand keep migrating throughout the year. During spring, birds migrate to temperate climates while come autumn they move again towards regions with warmer climates. The Arctic tern holds the longest migration ever recorded covering 49,700 miles. Arctic terns live for about thirty years. And in that period they migrate about 1.5 million miles – roughly equivalent to one-third the circumference of the sun (Zimmer, 2010). Bird migrations are primarily motivated by food supply. Birds usually migrate between hemispheres – switching hemispheres when conditions are not favorable. During winter, birds in the Northern hemisphere fly south and the same happens to birds in the Southern hemisphere.

Aircrafts use 'airways' to route their flight paths. Birds too have common flight paths they follow during their migrations – called 'flyways'. Aircrafts use satellite navigation systems as well as ground-based navigation transmitters (navaids) to navigate during flight. Using navaids they follow their route within airways. What about birds? How do they navigate through their flyways during migration? Scientists have been trying to sort out the answer since as early as the 18th century. During that time the Austrian physician, Franz Mesmer proposed the idea that inanimate objects could exert a force (like magnetism) on living animals, giving birth to the first theory of 'animal magnetism'. As is often the case with new theories, it soon became unpopular and gradually fell out of mainstream science. In the mid twentieth century, scientists started noticing certain peculiar behaviors of European robins in captivity. European robins have been known to migrate to the southern hemisphere during autumn. During autumn, these robins were seen to curiously gather at the southern end of their cages – it seemed like they were migrating within the confines of their cages.

In the 1960s, Wolfgang Wiltschko working at the J.W. Goethe University in Frankfurt experimented on these caged robins using electromagnetic coils. Electromagnetic coils are coils that become magnetic when supplied with electrical current. Wiltschko positioned the coils around the cage of the robins to alter the magnetic field. This caused the birds to gather at the other end of the cage. He discovered that he could influence the gathering of birds by manipulating the magnetic field. With the help of his wife, Roswitha, Wiltschko later experimentally proved that birds could indeed sense magnetic fields. And they use these fields to orient their movements (Meredith, 2014).

This was among the first experiments connecting avian movement to magnetism. Scientists today pinpoint bird magnetic sensing to two separate mechanisms. These magnetic receptors are thought to be inside the head. Before going further, let's digress a little and discuss the nature of magnetism. Magnetism occurs naturally in a mineral called *magnetite* found in iron ores. Certain materials composing iron, cobalt or nickel could also become magnets if they are magnetized. Magnetization is done by running electricity into the material inducing it to create its own permanent magnetic field. It then becomes a magnet (e.g., refrigerator magnets used to hold photos or notes on the refrigerator door). Magnets then have invisible magnetic lines which attract metallic objects towards it. Such lines are called a magnetic field. If the magnet is moved closer to another magnet with the same pole (north-north or south-south), then it will repel each other. Opposite poles on the other hand would attract both magnets. Now the earth can also be thought of as a huge magnet. Its magnetism is most probably caused by its metallic core. The magnetic poles of the earth are the inverse of its geographical poles. The geographical North Pole is the magnetic south pole and vice versa. Being a large magnet, the earth too has a magnetic field – invisible force lines originating from the geographic South Pole and terminating at the North Pole. Keep in mind the earth's magnetic field is extremely weak – at the equator it's about 160 times weaker than your average refrigerator magnet.

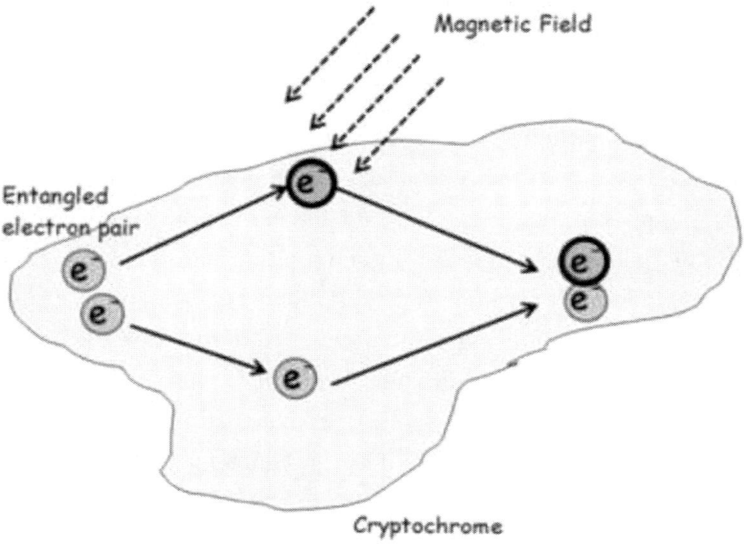

Figure 5.8. Cryptochrome Compass in Birds.

Returning to the two mechanisms birds use to sense magnetic fields. The first mechanism acts like a compass; so the birds use it to locate the direction of the poles. The other behaves like a magnetometer. A magnetometer is used by engineers and geologists to measure the intensity of magnetic fields. For instance, consider two magnets; labeled A and B. Let's say both magnets can attract a metal ball weighing 10 kg. When the 10 kg metal ball is replaced by one which is 15 kg, only Magnet B could successfully attract the ball towards it. This means that the magnetic field produced by Magnet B is more intense compared to Magnet A.

So birds have two in-built mechanisms for measuring magnetic fields (Wiltschko and Wiltschko, 2005). The Wiltschkos found that birds and certain other animals have small amounts of the mineral, magnetite in their bodies. When exposed to a magnetic field, the magnetite in the cells (especially at the bird's upper beak) gets affected. This alignment causes the cells to respond by sending signals to the brain resulting in the bird's movement. Magnetite in cells behaves like a magnetometer for birds. The earth's magnetic field varies depending on geographic features. For instance a mountain or hill may cause the field lines to spread out reducing intensity while a buried iron deposit may intensify the field. Therefore, using its magnetometer, birds build a mental map of the flight route so it can use it in later flights. They do this similarly to how people use landmarks to help them navigate when driving to a new location.

Second is the bird's compass mechanism. Unlike regular compasses, birds cannot tell which is the North and South Pole. Their in-built compasses let them know the direction where the equator and the poles are located. Previously, we discussed how the bird's entire body acts as a magnetometer. Interestingly this is not the case for the organ which acts like a compass. This organ is modestly located in both the bird's eyes. In fact scientists have narrowed it down to a single protein known as cryptochrome. To see how birds use cryptochrome to act as a compass, we must first understand a quantum entanglement.

Quantum particles can be connected or 'entangled' to each other. When this happens, tampering with one particle could somehow affect other particles. One way to effectively entangle particles is via supercooling – cooling them way below the material's freezing point. By doing this, all the particles get entangled. Imagine we take one entangled particle and carry it away to the other side of the world. Say we now tamper with this particle by exposing it to visible light (photons) – causing it to change state and become more energized. Then the entangled particles we left behind on the other side of the planet will change state as well. Disturb one and you disturb all. And this happens almost instantaneously. Einstein called this 'spooky action at a distance'. Since then quantum entanglement has eluded scientists; no one really understands how it exactly happens. Currently, quantum entanglement remains an active research area. So how does this relate to the bird's in-built compass?

In bird's eyes, electrons in cryptochrome proteins are naturally entangled. When a photon (light particle) hits the cryptochrome, these entangled electrons become separated by a very small distance – a few nanometers at most (see Figure 5.8). When these electrons are exposed to a magnetic field, they each change state – but in a slightly different way from the other due to the separation. And when they recombine in the cryptochrome, they do it differently depending on the direction of the magnetic field.

This change is then sent as a signal to the brain giving the bird a sense of direction, like a compass. Scientists believe that this reaction could give birds the ability to actually see the Earth's magnetic field in patterns of light and dark (Grossman, 2011). The research of Klaus Schulton and Thursten Ritz at the University of Illinois proved that the radio waves could disrupt the navigation of European robins. These works further support the avian entanglement theory (Ritz et al., 2000).

The biophysicist, Raf Freire suggested that even chicken may have magnetic receptors like many of their bird cousins. About a decade ago, Freire decided to run some tests to confirm his idea. He started by training chicks to search for a ping pong ball hidden behind a screen. The team trained them to align their search direction towards north where the ping pong ball is located. They then used copper coils to shift the magnetic field. This caused the chickens to change their direction of search according to the magnetic field. So the chicken's perception of north is totally dependent on the magnetic field. This was a simple and clear demonstration of how chickens use the Earth's magnetic field to navigate. The thing is, unlike their bird cousins, chickens don't migrate and hence have no need for this sort of sensing. It is possible though that chickens were not always as they are now. Perhaps earlier in their evolution before breaking off from the ancestral line of other birds, they may have used it to navigate during flight. Or even they may have used it to navigate on land like their close relatives; the junglefowl. What we are seeing now may be a remnant of its navigational past, as stated by Freire – "Jungle-dwelling chicken ancestors would have used it on their home range – about a square kilometer. It would be hard for them to distinguish trees in the jungle visually, so they would just use something as simple as the magnetic compass to navigate."

Besides birds, it has been recently discovered that humans too have cryptochrome in their eyes. In the eighties, Robin Baker of the University of Manchester tried to prove that humans can sense magnetic fields. He ran numerous experiments involving thousands of volunteers. In the end the results were inconclusive. And he never could pinpoint the exact mechanism at that time. Thanks to the research done on birds, we now know that cryptochrome is the basis of magnetic sensing.

Steven Reppert, a neuroscientist at University of Massachusetts decided to pick up where Baker left off. But this time instead of running arduous tests on humans like his predecessor, Baker took a genetic approach (Foley et al., 2011). Since fruit flies have cryptochrome, he genetically engineered the flies to become cryptochrome-deficient. When tested with a magnetic field, the absence of cryptochrome made the flies become poor navigators. He then inserted the human cryptochrome gene into fruit flies. Flies with the gene were seen to be able to navigate again. Thus the human cryptochrome protein is somehow linked to sensing magnetic fields. The funny thing is no human has ever consciously sensed a magnetic field. Most probably reading this you just came to know that you too can sense magnetic fields! Let's not get ahead of ourselves; truth is we don't really know how we use the cryptochrome in our eyes. In the case of humans, scientists speculate that cryptochrome may act more like a 'molecular clock' than a compass. Used in this manner it may function to regulate our circadian rhythm – which is a twenty four-hour biological cycle existing in most organisms. The circadian rhythm regulates the body's processes like sleeping and eating times, as well as more complex activities like hormone production and cell regeneration. Reppert also believes that cryptochrome may also aid in our visual-spatial perception – giving us the ability to gauge distance and depth accurately. Animals known to have magnetic sensing include: bees, snails, sea turtles, salamanders, newts, lobsters, butterflies, bats, salmon, frogs, trout, East African mole rat and whales.

Zooming-in on Evolution

Charles Darwin got the initial ideas of evolution theory in his twenties. He then busied himself for the next two decades stockpiling evidence to solidify the theory. Darwin knocked his theory around finding for weak spots – attempting to predict all possible doubts that may arise once it got published. His famed theory sits on the concept of natural selection. Natural selection occurs when two individual organisms with different genes produce an offspring – which matures and then repeats the process. As this continues only genetic traits which are advantageous are retained (or naturally selected) in the population while the rest are discarded. This improves the fitness of the subsequent generations increasing its chances of survival. The catch is: the whole thing only works if there is genetic variation in the population. Without variation, the same genes would keep turning up and gene selection will get very limited. So after say a few generations, the individual with the exact same genetic make-up like their grandparents (or some ancestor) may be produced in the population. Without genetic variation, the population will reach its maximum fitness and stay there – its adaptability will come to a halt. Any further changes in the environment would wipe out the entire population.

Mutation plays an important part in maintaining genetic variation as it gets passed down from one generation to the next. This keeps the *gene pool* diverse allowing the population to keep evolving. Gene pools are simply all the genes in a given population. Besides mutation, gene pools get diversified if the individuals in a particular population reproduces with those from another. This cross reproduction increases the richness of the gene pool by the acquisition of new genes. The acquisition of genes is called *gene flow*, since the genes flow from one population to another. For large animals, gene flow happens slowly over the years. This is not the case for microbes. Their fast metabolism and short life span allows for rapid reproduction – running through generations in a matter of hours.

If a population of organisms does not interact with other alien populations, then gene flow does not occur. This means that the population will completely depend on mutation for genetic diversity – known as *genetic drift*. When discussing genetics of populations, it is hard to avoid mentioning the Galapagos Islands – the place that first inspired Charles Darwin to propose the theory of evolution while aboard the H.M.S. Beagle. The Galapagos consists of a series of volcanic islands along the equator in the Pacific Ocean. Volcanic islands come to existence from volcanic activities under the sea which then erupt above the water surface forming a land mass. Animals and plants on such islands usually arrive from elsewhere – perhaps deposited by migratory birds or transported by drift wood from other places.

The Galapagos is home for many fascinating creatures such as the land and marine iguanas. No one knows exactly when the first iguanas arrived at the Galapagos. Our best bet is that they arrived from the mainland to one of the islands. As the islands sunk they migrated across to other islands. The crossings would have been easier in the beginning when the islands were nearby. The iguanas could easily hang on to driftwood and get washed to shore on other islands. But with time islands are known to move far apart due to geological effects. This is most probably when the iguanas got isolated from other similar iguanas. These isolated populations would then have a limited gene pool causing genetic drift. As time passes, these iguanas evolved independently drifting apart from others of their kind. Prolonged genetic drift eventually results in a new species (*speciation*). After evolving into a new species, these iguanas lose their capability to breed with the other iguanas which are now genetically too distinct from themselves. This is why today we have three different species of land iguanas in the Galapagos. On one of the islands, these iguanas evolved to the extent that instead of grazing plants on land they started feeding on seaweed. Thanks to natural selection, they quickly adapted to the sea becoming expert swimmers. Marine iguanas have evolved from the land iguanas but not from the ones existing today. We know that marine iguanas being only native to the Galapagos may have come from a family of extinct land iguanas (Dawkins and McKean, 2012).

Marine iguanas and other land iguanas could not have evolved in isolation without mutation. In the absence of mutation and gene flow from other colonies, their gene pools would be too limited causing them to lose adaptability and survival fitness. This would eventually lead to extinction. This is how important mutation is for the survival of life forms. But what exactly is mutation and how does it work? Evolution is often referred to as the work of a 'blind watchmaker'. This analogy originates from the random nature of mutation. During natural selection, an organism randomly mutates to predict environmental changes. Thus when the changes occur, the organisms with the right mutation survives while the ones with less advantageous mutations perish. Mutation could be thought of as a genetic betting game – where the genome randomly places bets on potentially useful traits for adaptation. This way evolution, the blind watchmaker generates complexity simply by a series of random changes. Even one of our initial quantum biologists, Max Delbrück ran groundbreaking experiments in the early forties strengthening the blind watchmaker analogy. Delbrück and his colleagues showed that spontaneous mutations occur in bacteria populations when exposed to environmental pressures. They explained that mutations randomly happen in the genome but they are very slight. Over time these mutations accumulate significantly influencing the organism's adaptability. This was the work that won Delbrück the Nobel Prize in Physiology. As I mentioned initially in this chapter, he was also a collaborator and friend to a prominent physicist, Wolfgang Pauli. In light of Delbrück's experiments, Pauli did have doubts if mutation was truly random. In correspondence with Delbrück, Pauli was skeptical if 'a simple probabilistic model would not be sufficient to generate the fantastic diversity we see'. Thanks to the development of powerful instruments, we now roughly know something about the mechanism of mutation. Mutations happen when the DNA is being copied or repaired. During this, the molecular machine (enzyme) copies the wrong code or inserts the right code in a wrong place on the genome. This may change how the DNA is being read to produce proteins – this influences the organism's adaptive traits.

After Pauli, a series of experiments in the late eighties brought the blind watchmaker analogy to doubt again. This time it was by John Cairns and colleagues working at Harvard University. His results showed that bacteria could somehow choose which genes to mutate and thus control its mutation rate. The experiments involved bacteria which are incapable of digesting a kind of sugar called lactose. So they placed the bacteria in a lactose-rich environment. With no other food source available, the bacteria mutated and started consuming lactose. Under duress, it mutated. Previously it has been known that certain cells in the body mutate at a higher rate compared to others. For instance, the cells in the immune system could mutate rapidly producing antibodies to fend off infections. However, their mutations are sort of controlled since they cannot pass on the mutated genes to the next generation of immune cells. This may be some kind of biological fail-safe mechanism so that they don't uncontrollably mutate and attack the organism's own cells – creating what is called an autoimmune reaction. After Cairns, the research was continued by Susan Rosenberg from the University of Alberta and Partricia Foster working at the University of Indiana, Bloomington. Running experiments on the genome of the bacteria, *E.Coli*, they found that the bacteria could not choose the mutated genes. So Cairns' hypothesis was partially wrong since Rosenberg and Foster showed that the mutations were spread across the entire genome. They still observed what Cairns had initially seen – that environmental pressure caused the bacteria to increase its mutation rate. Trying to understand how this was happening, Rosenberg looked closely at the molecular aspect of mutation. She noticed that under stress, the enzyme that regularly copies the DNA is replaced with a 'mistake-prone' enzyme. This new enzyme goes around mistakenly copying the DNA generating mutations in the bacteria.

Right around the same time another similar research was being carried out by physician, Peter Glazer at Yale University's School of Medicine. But Glaze was not looking at bacteria; he was more interested in cancer cells. Glazer found his results consistent with Rosenberg and Cairns' experiments. Rapid growth of tumors tend to put cells under pressure by restricting blood flow, food and oxygen. When this happens, the cells increase their rate of mutation becoming cancerous. Similar to the mechanism proposed by Rosenberg, Glazer states that during this time, the cell's ordinary DNA repair mechanism is halted. Bear in mind this whole business of mutation rates caused a great stir in the scientific community. Letters were shot back forth and rebuttals were written countering each other's arguments. At the time this book is written many of these arguments are not yet settled. I will not go into the details related to these exchanges in this book (Singer, 2014; Rosenberg et al., 2012; Foster, 2004). Suffice to say that the findings of Cairns, Rosenberg and Glazer were far reaching, echoing across various corners of the scientific community.

The mutation experiments by Cairns piqued the interest of the biologist, John Joe McFadden and physicist, Jim al-Khalili at the University of Surrey. They looked up an older research work published by Per-lov Löwdin, a Swedish physicist. In that work, Löwdin proposed a mechanism called proton tunneling in the DNA as a cause for mutation. Building on that work, McFadden and al-Khalili drew up a quantum picture as an explanation, here it goes. The double helical structure of the DNA is held together by hydrogen atoms (see Figure 5.9). The hydrogen atom connects the bases on both sides acting like a rung in a ladder. Bases are the codes making up the chromosome as shown in Figures 3.1 and 3.2 in Chapter 3. Remember, all this is happening at a quantum scale, so we can expect some quirky physics. Usually, the hydrogen atom sits closer to one-side of the DNA ladder compared to the other. Mutation happens when the hydrogen atom tunnels to the wrong side of the DNA ladder (Merali, 2014).

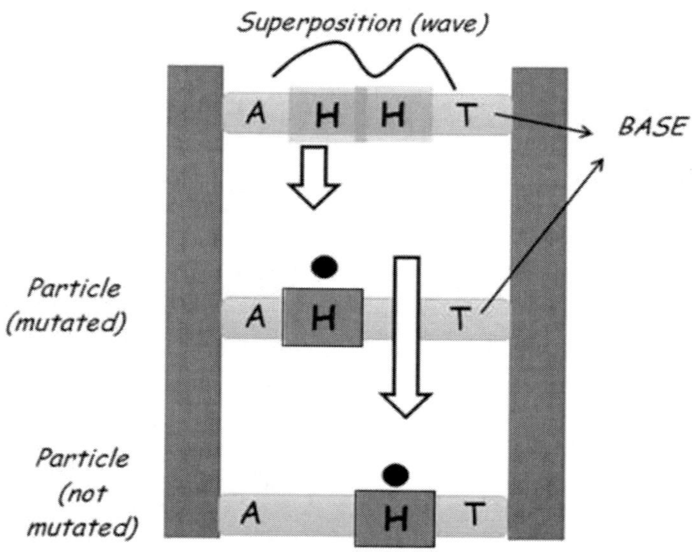

Figure 5.9. Ladder Representation of Hydrogen Tunneling in DNA.

Let's recall Turin's quantum scent theory. Before the odor molecule arrives, the electron at the edge of nose receptors are in a wave form – maintaining superposition and thus being all over the receptor at once. The vibration of the odor molecule observes the electron and thus causes it to tunnel through the barrier reaching to the other edge of the receptor. This is how it collapses into a particle. Something similar happens to the hydrogen atoms in the DNA. Only this time the environment acts as the 'observer' instead of the odor molecule. Consider the lactose-intolerant bacteria in Cairns' experiment. During stress, the lactose from the environment observes the hydrogen atom in the bacterial DNA. Before this happens, the hydrogen atom is in a state of superposition – mutating and not mutating the DNA at the same time. Therefore, is sits at both edges of the rung. When the lactose from the environment makes the observation, it then collapses into a particle – tunneling towards one edge of the rung inducing a mutation. If the hydrogen tunneling theory is right, then mutations may not be as random as we thought they were. They may very well be environmentally influenced and powered by quantum mechanics.

TINKERING WITH LIFE

The delicate quantum mechanisms described in the previous section may seem a little extravagant or even outlandish. To think that nature could come up with such complexity and detail does astonish even the best of scientists. Come to think of it we even have problems reproducing many of these biological phenomena in labs. The truth is; we barely scrape the surface of the natural world with our understanding of science as it is. We have been at it for say a couple of hundred years at most. Now the earth on the other hand is about 4.5 billion years old. And that is a long time for nature to run its experiments on the evolution of life. It can even afford to keep failing – improving slowly as it goes along until it figures out the correct physics required to perfect the sensory needed for smelling or the hydrogen tunneling framework needed for mutation. Time is the key element here; nature's secret weapon. Science offers us the means to accelerate certain natural processes so we can get some idea of how it works. For instance we can recreate chemicals in mere minutes which would take years to naturally occur. Computer simulations can recreate millions of years of earth's climatic and geological changes.

Fossil evidence indicates that nature was somehow experimenting with life in its early stages. We encountered such fossils when we discovered the Burgess Shale located in Canada. The fossils found there dated back to the Cambrian period – which was about 541 million years ago. The Cambrian period is part of a classification method used by scientist to easily distinguish different periods in the history of life on Earth. Fossil records show that during this period there was a sudden rise in animal life – this event is referred as the 'Cambrian Explosion'. The event was followed by major diversification of life on the planet. Many major groups of animals we have today could be traced back to the Cambrian Explosion – including humans. The Burgess Shale in the Canadian Rockies was first discovered by Charles D. Walcott of the Smithsonian Institution in the year 1909. Few legends exist on how the discovery was made. According to one of them, the horse ridden by Helena, Walcott's wife slipped overturning a rock (Adler, 2013). This rock revealed the first of the specimens found there. Upon this unexpected discovery, Walcott knew he was onto something. So he returned the following year and collected 65,000 specimens of about 127 species.

Most of the specimens he found have been seen before in previous excavations. These specimens were mainly ancient arthropods – creatures with exoskeletons similar to insects we have today. Many of the insects found today like spiders, scorpions and centipedes have been known to evolve from those older arthropods. Among these ancient arthropods is a marine creature called trilobites. Found in various excavation sites around the world, trilobites have bodies spanning from about 1milimetre to half a meter in length. And they were all plated with body armor. The armor was most likely used to protect them from other sea predators. Plenty of trilobites were also found at the Burgess shale. Compared with other excavations, the Burgess Shale contains the best record of animals from the Cambrian period. This is because the animal specimens found there were very well preserved within its layers. In most cases, fossilization happens when animals die and get buried under layers of sediments. This happens especially if it dies in a river or a lake where sediments slowly cover the body – and not completely avoiding decay. On the other hand the animals found at the Burgess shale were aquatic and living in underwater mud banks. From time to time the water currents would cause a mudslide rapidly burying the animals with sediments. This speedy burial allowed for preservation to begin immediately. This is why animal fossils found at the Burgess Shale are very well preserved.

Walcott's collection also included strange animal specimens never seen before. One example is the five-eyed creature, *Opabinia.* Being a predator, this creature catches its prey using a clawed arm protruding from its head. When the creature was first presented at a scientific conference, the audience mistook it for a practical joke. Another curiosity found there was a marine worm named *Hallucigenia.* Due to its confusing features, this worm was first constructed upside down and front to back – which explains the weird name. *Hallucigenia* worms had a tube-like body with seven or eight thin clawed legs. Above each leg at the back of the worm contains a cone-shaped spine. Clearly this is not your regular garden variety worm – this one is many-legged and has claws. Adding to that it has a protective spine probably to fend against predators. Another Cambrian marine predator which is truly the stuff of nightmares is the *Anomalocaris*. Growing up to two meters in length, this animal looks like a huge shrimp. Its mouth underneath its head is lined with hooks used to trap prey. Closing like nippers its mouth is believed to crush hard-shelled arthropod prey, like trilobites. Among this menagerie of curiosities, scientists also found a worm-like creature, *Pikaia* which have seem to have grown a rod-like primitive backbone. This was most probably the ancestor of backboned animals (vertebrates) 'we see today.

These strange animals were part of the sudden proliferation of life during the Cambrian period. From archeological evidence, we know that nothing like these specimens have ever turned up in other sites dating back to the Cambrian era. Walcott tried although without success to relate these specimens to other known Cambrian era creatures. They were so strange that he could not even classify them to any living descendants seen today. A few decades later, the geologist Harry Whittington and his team from Cambridge University had another go at the problem. They then realized that these specimens were so unique that they belonged to a completely new phylum. A phyla is one of the most general classification (taxonomy) used to categorize living organisms. Let's take humans as an example (see Figure 5.10).

Our species is *Homo sapiens* while our genus is grouped as *homo*. Currently our subspecies is called *Homo sapiens sapiens* to differentiate as with our direct ancestors *Homo sapiens idaltu*. We belong to the *hominidae* family which other members include the great apes (chimps, gorillas and orangutans). Generalizing further we reach the classification of order. We are in the order of *primates* along with many other monkeys. Going up the ranks we are now at category of class. Humans are in the class of *mamalia* (mammals). Finally we get to phylum at which we are grouped as *chordata*. The phylum *chordata* consists of animals with spinal nerves and tails occurring anytime during their lifecycle. Wait a minute, but humans don't have tails? This not entirely true since our tails form for a short period when we are still embryos. The tail eventually disappears but remains prominent when the human embryo is between 31 and 35 days old (Fallon et al., 1978). Generalizing further, above phylum we have kingdom and above that we have the category of domain. We belong to the *Animalia* kingdom and the *eukaryota* domain.

This is what the Cambridge geologist, Whittington realized – some specimens found at the Burgess Shale require us to define a whole new phylum to account for their existence. Therein lies the question: what happened to the descendants of this phylum? We know that those animals descended from the *chordata* phylum are present today – humans and many other invertebrates (animals with spines). It is possible that animals from this phylum went extinct due to evolutionary selection. From the fossils one thing is clear – many of the animal specimens living there were carnivorous. With very few plant-eating animals (herbivores), bigger animals preyed on smaller ones. This could be seen from the various protections used by smaller vulnerable animals – spikes, body armor and thick shells. The predators on the other hand developed claws and jaws meant for crushing shells and keen eyes for spotting camouflaged prey. In other words, the Burgess Shale animals were engaged in an evolutionary arm race. It may be that the only animals which survived this arm race reproduced successfully – passing down their genes to subsequent generations over the years. Those that lost the competition perished.

Perhaps the Cambrian Period was a time when nature experimented with life. It gambled by producing a very diverse range of animals with all kinds of weird features (five eyes and hooked mouths). Due to survival competition, only the ones with advantageous features survived. So the failed subjects became extinct thanks to selection. The descendants of the surviving animals are probably the ones we see today. In his book, the paleontologist, Stephen Jay Gould wrote that survival selection of animals during the Cambrian period could be likened to a lottery (Gould, 1990). If the clocks were turned back, the outcome may have been quite different. We may find different animals currently roaming the earth. This is of course assuming that our previous ancestral lineage (*Chordata*) managed to thrive in the first place so we can exist today.

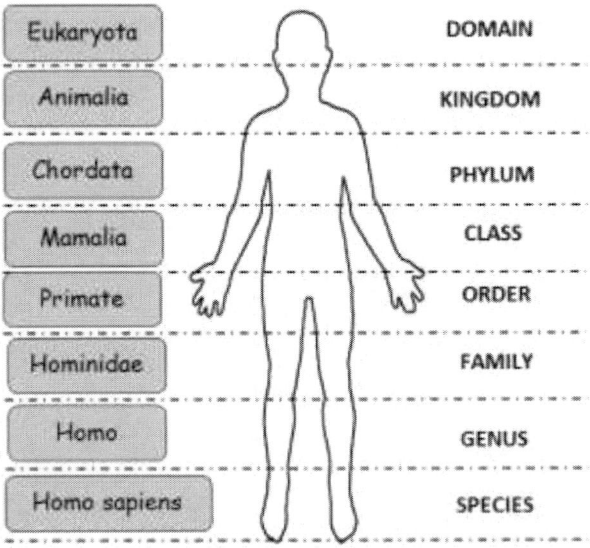

Figure 5.10. Human Taxonomy.

Turning Back the Clocks

Imagine how things would be if the conditions of the Cambrian period were slightly different – where a certain species won the evolutionary battle instead of another. To know the impact of this to current life on earth would be quiet interesting. The two prevalent concepts in evolution are convergence and contingency. Convergence, also known as convergent evolution is a scenario where different organisms eventually end up having the same features. A good example of this is echolocation; the ability to use sound for navigating (sonar). Dolphins, porpoises, bats and shrews have all developed the ability to echolocate. Notice that these animals are from completely different species (or even order) and yet they all developed the same feature; echolocation (Parker et al., 2013). Another example is red blood cells. Almost all backboned animals (vertebrates) evolved red blood cells with the exception of certain salamanders and fishes (Hoffman et al., 2010). Bone structure in vertebrates can also be very similar suggesting convergent evolution. Figure 5.11 shows the similarity between the limb structure of a human, dog, bird and whale.

The idea here is that no matter what the initial conditions or the selections during evolution; the outcome is always the same. Organisms with certain features will always prevail. In this view, irrespective to the past, humans with our intelligence will eventually come to existence. Some mammals like dolphins and whales would become aquatic sooner or later while certain reptiles would lose their legs and slither around becoming snakes. This is the essence of convergent evolution. So the similarities in features we see among different animals are just evidence indicating that these are the features that would eventually prevail. These special features are like 'rigged lottery tickets' designed to win the evolutionary game. So if you happen to be an organism which doesn't have the winning tickets, then you are done for. You are heading right on the path towards extinction.

The opposing view of convergent evolution is contingency. According to this approach, the outcome of evolution is very sensitive to initial conditions and the results of competition during the process. This means that if conditions were slightly different during the Cambrian period, certain reptiles may not come to being. Perhaps even humans may not have existed. This is something like chaos theory which we visited in the last chapter. There we saw how small differences in the beginning could result in huge impacts – show by the population of cats in Figure 4.1. The convergent evolution view is more rigid and deterministic as compared to the contingency approach. But which is right? Convergence or contingency? So to test this out, the evolutionary biologist Richard Lenski ran some experiments to see if he could catch contingency or convergence in action (Zorich, 2016; Sniegowski et al., 1997).

The problem with evolution experiments is that we cannot simply turn back in time and see what happens at certain periods. Real-life is not a video player; we cannot simply scroll back and forth in time to observe some particular event. What we can do is to try and capture these events as we inevitable move forward in time. In fact, these days we almost routinely do this and publish it online via mobile applications like Snapchat and Instagram. By taking snapshots as we go forward in time, we capture frozen moments in the past. So we can always go back to different instances in the past and review them. This is what Lenski did. But instead of using a camera, he used a freezer to take snapshots of evolving bacteria. Working in the late eighties at Michigan State University, Lenski used the bacteria *E.Coli* as test subjects. He separated the bacteria from a single population and placed them into 12 different flasks containing nutrients (bacteria food). Then every few months for the next twenty over years, his team froze one of the bacteria-filled flasks. By freezing it he can capture the bacteria's state of evolution at that point in time. So he froze the flasks every few months accumulating various frozen states of bacterial evolution for the past twenty something years. A pretty neat trick!

By thawing out the bacteria he can then run tests to see the physical and genetic changes in bacteria at various points of its evolution. In 11 of the flasks, most *E.Coli* grew larger. In one of the flask, something strange happened; the bacteria seemed to split into two groups. One group became larger and the other smaller. They somehow evolved to live together in a single flask. This is like the situation in H.G. Wells' classic novel, The

Time Machine – where in the year AD 802,701, humans evolved into two separate species (Eloi and Morlocks). Lenski reported that both bacteria groups coexisted in the same flask for 50, 000 generations. This is contingency number one. The second happened in 2003 where one of the flasks became murky. At first the research team thought that the sample in the flask was contaminated. Upon further tests, they realized that the bacteria in the flask have evolved to digest another chemical, citrate besides their usual food (glucose). This is similar to what happened in the John Cairns experiments at Harvard which we encountered when discussing the quantum nature of mutation. In those experiments the bacteria evolved to consume lactose due to the unavailability of its usual diet. Although similar, this is not exactly the case here. The *E.Coli* in Lanski's experiment were not under any kind of duress – the nutrients in the flask had many different chemicals. His bacteria simply evolved to optimize its food consumption by ingesting citrate in addition to glucose. In the past 15 years which is about 31,500 bacterial generations, only one of them evolved to ingest citrate.

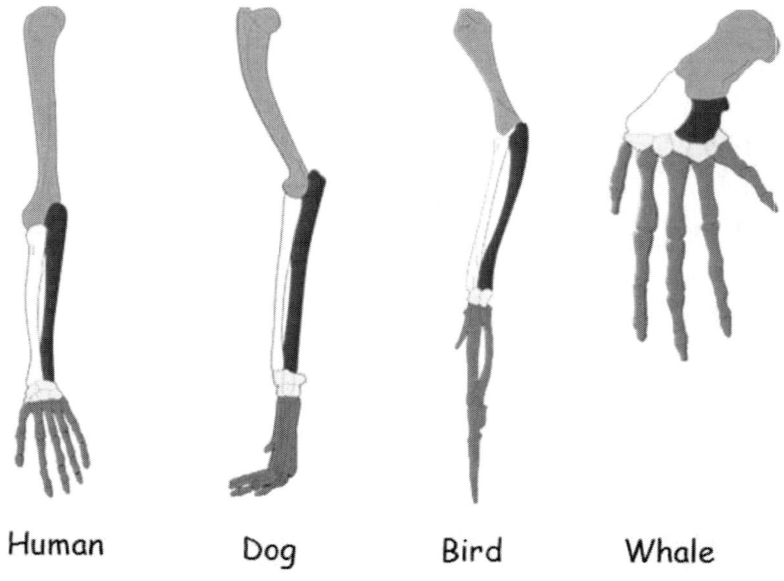

Figure 5.11. Skeletal Structure similarity (Source: Palaeontology [online] (2017)).

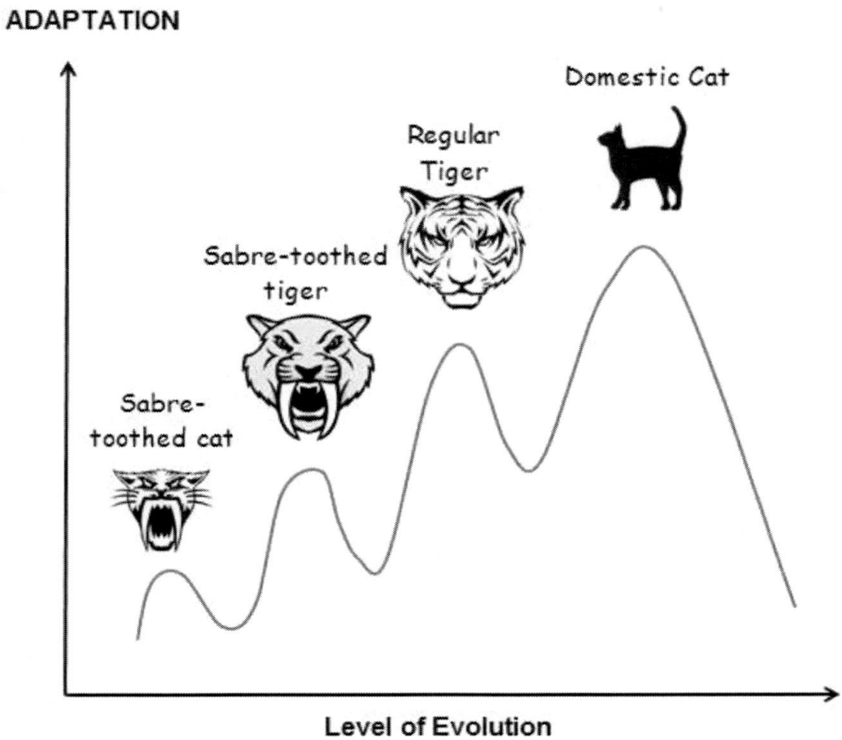

Figure 5.12. Adaptive Peaks.

His team also found that mutations were not a single event. For instance, let's say humans like bacteria evolved to consume certain chemicals like plastic. This sudden change would not be happening from a one-time mutation. But rather a series of mutations in the past. The mutations compound creating the potential until the time comes to completely switch diets and consume plastic. So if there is a change in the history of the mutations, eating plastic is out of the picture. Perhaps we would develop some other mutation – like becoming more vegetarian? So accumulation of the types of mutations is what counts. Anyway, the Lenski experiments are all for contingency.

At Harvard University, David Liu and his colleagues managed to discover another way to run evolution experiments. Lenski's snapshot approach only works because the entire evolution of his bacteria, *E.Coli* took decades. But what if you could cut the evolution time making decades look like hours. This was the approach Liu and his team employed. They found a way to speed-up evolution by increasing mutation rates. They did this on a certain type of virus that attack only bacteria – phage. They called this technique PACE, which stands for Phage-Assisted Continuous Evolution (Packer et al., 2015). Using PACE, Liu and his colleagues first separated the viruses into different compartments. By maintaining the same conditions in all the compartments, they then manipulated the rate of evolution. The team pressured the viruses to produce a certain enzyme to observe how they evolve. For simplicity, let's call it Enzyme A. After some amount of evolution, the viruses that could not produce Enzyme A were removed from further tests.

Bear in mind these are all the same viruses in the beginning. The only factor varying is the rate of evolution. So the viruses evolve differently, some faster while others are slower. These differences give rise to different strains of virus. Let's consider an extreme example shown in Figure 5.12. Imagine first we have a prehistoric saber-toothed cat. Then we apply a similar technique which speeds its evolution. Let's fix the time to say one hour. Just like Liu we apply different evolution rates. At one evolution rate, we would get a saber-toothed tiger after one hour. By using a higher evolution rate, we then get a modern tiger in an hour. Then we use an even higher evolution rate, only now we have the usual domestic cat. So at different evolution rates we get different strains popping out after an hour using the same initial subject, the saber-toothed cat. This is simply an exaggerated version of evolution which is used as an example here. The evolutionary progression of domestic cats is not that straightforward.

In the end Liu and his team obtained a series of evolved virus strains that produced Enzyme A. But the effectiveness of the enzymes they produced was different depending on the virus strain. Some strains produced highly effective versions of Enzyme A while others produced lesser ones. Liu's experiments showed that all the viruses evolved to produce Enzyme A. But they also got stuck at some point – where some viruses kept producing a certain type of Enzyme A and could not evolve further to produce better versions.

This is like saying that in our saber-toothed cat example, some cats will remain as tigers. No matter how much evolution it undergoes, it sort of gets stuck – not capable of evolving further to become a domestic cat. Some viral strains seem to hit an evolutionary brick wall which halts their progress.

Their results show that there are various degrees of adaptation. These levels are explained by 'adaptive peaks' (depicted in Figure 5.12). Consider the first saber-toothed cat (Peak 0). This time let's not play with mutation rates and leave the cats to evolve naturally instead. Given enough time the earth's temperature got colder so these saber-toothed cats got bigger and evolved to saber-toothed tigers (Peak 1). Since the prey got smaller they didn't have use for the saber-teeth and hence lost it becoming a just tigers (Peak 2). As more time passed, food became abundant since it was provided to them by their human caregivers. The tigers then shrank and became domestic cats (Peak 3). This is the optimal evolution path. However, in some cases saber-toothed cats did become bigger during cold temperatures but not big enough to become a sabre-tooth tiger. So it evolved inefficiently. Since it did not become a sabre-tooth tiger, its consequent evolution to become a regular tiger gets affected. Not achieving Peak 1 makes it difficult to climb to Peak 2, since it would have to de-evolve itself back to Peak 0 and start back again. The same goes for the following peaks resulting in a diversion from its evolutionary path – from the optimal route. The key elements here is timing and adaptation efficiency. If you start on the wrong foot, the subsequent adaptations will vary as well. This is why the notion that a single mutation resulting from a single adaptation does not accurately describe evolution. It's more like a series of mutations prepare the organism for subsequent mutations. These initial mutations are crucial as it builds the potential for the coming evolution. Therefore, a saber-toothed cat cannot jump across its adaptive peaks to become a domestic cat. It would have to cross all the peaks becoming a saber-toothed tiger and then a tiger before it reaches the domestic cat. And if along the way it does not make it to one of the peaks for some reason, then its evolution may get halted. Not being able to climb the next peak, its evolution may take another path. As with Lenski's experiments, David Liu and his team took the side of contingency. Their results indicate that if we were to turn back the clocks, evolution would most likely take a different route. Due to the randomness of initial conditions and sensitivity to mutation, the exact evolutionary path would unlikely repeat. So if the Cambrian period had a different evolutionary outcome, then life on our planet would definitely differ and our very existence will come to doubt as well.

Some researchers ran evolution experiments virtually using genetic algorithms (see Figure 4.7 in Chapter 4). The computational biologists, Charles Ofria and Chris Adami developed a computer program, Avida for this purpose (Ofria et al., 2005). They created artificial organisms with digital genes evolving in a virtual environment. They evolve undergoing mutation, cross-over and selection exchanging genes as they go on. The researchers evolved the organisms to solve math problems. They realized that the successful organisms were those that had many seemingly bad mutations in the beginning. These mutations seemed to create the potential for advantageous mutations in the future. Such initial mutations diversify their virtual genes enough so that the subsequent mutations could occur with ease. Their results sound similar to the experiments done by Lenski and Liu in real-life.

How about findings suggesting convergence? One research along this line was carried out by physicist, Michael Desai at Harvard. Using robots he managed to run large-scale evolution experiments on yeast (the fungus used for baking bread). He found that the evolution was indeed sensitive to initial mutations (Kryazhimskiy et al., 2014). But he also saw that at very large scales, yeast tend to converge towards a same point no matter which path it takes during its evolution. This means that the adaptive peaks it passes does not make a difference to its overall evolution. We may have to take this results with a pinch of salt because these tests were run only using yeast. It's possible that yeast does not have many evolutionary peaks or its peaks are all close-by making them seem similar – giving an illusion of convergence. Additionally, over very large scales certain similarities do recur as we have seen in Figure 5.11. This still means that evolution would not repeat if we turn back the clocks. Perhaps they may be some similarities among the organisms but contingency may still dominate ironing out convergence. This is of course conjecture on my part. We have yet to uncover the interplay between convergence and contingency in evolution. Further investigation is definitely required to answer the question: Will life be the same if we turn back the clocks?

MORE CURIOSITIES

Having a combination of mammalian and reptilian features, the platypus stands as an evolutionary enigma – a true symbol of animal diversity fitted into a single species. From fossil evidence we managed to learn a couple of things about its evolutionary past. We know that platypus' split from the hereditary line of birds and reptiles about 280 million years go. After the split they became mammal-like reptiles. Then at about 150 million years back they split again from that line becoming completely mammal. Nevertheless till this day, they still held on to some of their reptilian characteristics. Although we only find them today in Australia, fossils of these creatures have been identified in South America. Identifying the animals from fossils is relatively simple thanks to their give-away beak and their odd bodies. Among mammals on earth, two have been known to have undergone minimal evolutionary changes – platypuses and opossums. These were the findings of Samantha Hopkins, a geologist working at the University of Oregon (Langley, 2015 and Langley, 2016). Looking at the evidence, scientists believe that the oldest fossils looking like the modern platypus dates back to about 2.5 million years ago. Possums on the other hand go much further. They haven't changed much for the last few tens of millions of years. Possums seemed to have split the lineage of marsupials (e.g., kangaroos, wombats, wallabies, etc.) about 60 million years ago (Horovitz et al., 2009). Found mostly in Tasmania, New Zealand, Australia, New Guinea and the Americas, from then on they hardly changed.

Though the platypus didn't evolve much in terms of its features, it did, however, shrink in size in the last 10 million years or so. Fossils of a much bigger species of platypus have been discovered in Queensland, Australia indicating that the earlier species was about the size of a domestic cat (Pain et al., 2013). The reason for this shrink in size is not clear, although scientists speculate that the earlier platypus species may have been carnivores that adapted for bigger prey such as frogs compared to modern platypuses – which only consume small shrimps, worms and insect larvae. Platypuses have certain features that mark them right off in the book of mammals. Among them is electrolocation. Previously we discussed how certain animals like bats and dolphins evolved echolocation – navigating using sound waves. Birds on the other hand could navigate using magnetic fields. Platypuses navigate using electricity. This means that evolution has used any feasible physics for developing sensing and navigational means in animals:

- Sound waves – echolocation and normal hearing.
- Magnetism – avian navigation, human circadian rhythm.
- Electromagnetic waves (visible light) – eyesight.
- Molecular vibration – smell.
- Electrolocation – platypus, cockroaches, Guiana dolphin, sharks and echidna.

Aquatic animals have been known to have electrical receptors used for navigating as well as locating prey. The only mammals which can electrolocate are platypuses and echidnas. The echidna is a type of anteater sometimes known as the spiny anteater. These are the only two mammals which lay eggs. Found mostly in New Guinea and Australia, echidnas fall into the same order as platypuses. Platypuses have about 40, 000 electroreceptors at their beaks. On top of that platypuses could also give a nasty sting if intimidated. It injects venom using sharp stingers located on the heels of its rear feet. Although not deadly to humans, the venom containing 83 known toxins could cause severe pain. Further analysis has shown that platypus' venom gene is similar to that found in the DNA of spiders, snakes and starfishes (Whittington et al., 2010). Another strange feature of the platypus is that although they have mammary glands which produce milk for their young, they don't have nipples like other mammals. Thus the platypus releases milk through its skin accumulating it at certain groves near the abdomen. The young feed by lapping up the milk at that region. Although the platypus is aquatic, it never evolved the ability to hold its breath for longer periods submerged. It can only stay submerged for no longer than half a minute.

As reported by the geologist Samantha Hopkins, the platypus has evolved much in the past couple of millions years. Even how it got to its current state remains an open question. Perhaps the platypus was in the midst of evolving when the conditions suddenly became stable. Imagine millions of years back there was a sharp variation in climate causing changes in the ecosystem. This in effect shifts the availability of prey and resources for the platypus. So it happily evolved. But in the midst of its evolution, conditions suddenly stabilized. Prey was abundant and the climate settled. This way its evolution came to a sudden halt – freezing it in its current state, partially evolved. Plus the platypus does not really face any serious competition from other Australian aquatic organisms. So it doesn't need to evolve.

Elephants on the contrary have evolved extensively over the years. Originating in Africa, they migrated across to the vast regions of Europe and Asia. Along with the migrations came evolution. Like the Galapagos iguanas, the elephants started to break into a few species; Indian elephant, Asian elephant, African Bush Elephant, Borneo Elephant, etc. Even before their great migrations, these animals evolved trunks – a useful thing to have if you are short in size and have to reach high up the trees for food. One problem solved. But the trunk takes up a lot of room on the head – so its skull changed shape and its incisors became tusks. Two down. With the tusks and the trunk, there's little room in the mouth for the teeth. So the teeth have to change too. Due to the shortness of the jaw, the elephant could not have a full set of teeth like most other mammals. So elephants evolved a unique set of teeth rarely seen in other members of the animal kingdom. Instead of having a series of teeth, it evolved to have a single long molar in its upper and lower jaw (Robert, 2016). Technically the elephant has four teeth; a single tooth for each side of the upper and lower jaw. Then the elephant for reasons unknown switched its diet from soft leaves to rough grass – at about 8 million years ago. Most probably the abrasiveness of the new diet caused its teeth regeneration to evolve (Lister, 2013; Saarinen et al., 2015). As the elephant's teeth wears down, it regrows its next set. An elephant can regrow its teeth as much as six times during its lifetime. It may be that the environment and availability of resources caused the elephant to change its diet, just as it made the platypus to become venomous. Or could it be that it changed its own behavior first before influence from the environment? For instance, perhaps the platypus suddenly preferred living in holes of snakes. And to chase the snake out of their original homes it developed venomous stingers for offense. In the case of the elephants, perhaps it developed a taste for abrasive grass rather than soft leaves. Hence it evolved regenerative teeth. Evolution could work in both ways; where changes could be from either external pressures or from the organism itself. It would be interesting to find out more about this aspect of evolution – changes induced by the organism itself. What would cause an organism to suddenly change its lifestyle so much so that it forces the hand of evolution? Maybe a change in genetic makeup resulting from a viral infection? Or perhaps it's more ecological; some indirect interaction with an organism in its ecosystem – like a parasite? Only more research can shed light on these questions.

Rise, Red as Dawn

Environmental conditions do influence the evolution of organisms. Besides geological and climatic elements, the key factors impacting evolution are availability of food, competition with other species and potential threats (from parasites or predators). So in many ways evolution of a particular organism is very much dependent on the organisms it interacts with. Consider the evolution of a cheetah. The maximum velocity and duration a cheetah could run would depend on the running capability of its prey, the gazelle. Having abundant food supply, the cheetah would multiply and more gazelles would be hunted. Since they always catch the gazelle, the cheetahs would eventually exhaust their food supply causing the gazelles to go extinct. What if the reverse should happen; where the gazelles could always outrun the cheetah? Then cheetahs would have to find for alternative prey. And if this fails, the cheetah would be the one getting wiped out this time – from starvation. Let's say the gazelles genetically mutate to run faster. The cheetahs would respond accordingly by evolving stronger leg muscles to catch up to the gazelles' speed. The reverse could happen as well, where the gazelle evolves to keep up with the changes in the cheetah.

So organisms co-evolve with each other to maintain their stable interaction. They keep changing to each other to remain with each other or to just simply survive. This is why co-evolution is also called the Red Queen Effect – coined by the American biologist, Leigh Maiorana Van Valen. That's right, as in the Red Queen in Lewis Carroll's 'Through the Looking-Glass'. In the novel the Red Queen says to Alice, "Now, here, you see, it takes all the running you can do, to keep in the same place." The change in the environment causes direct changes in certain organisms. Others linked to these organisms are thus forced to adapt as well creating something like a chain of adaptations. In an evolutionary context, it takes all the evolving one can manage to simply survive. The Red Queen Effect reminds us that evolution is a dynamic process – such that life on the planet is in a state of constant change.

In some cases the Red Queen Effect can be beneficial to both organisms. An interesting example is the evolution of hummingbirds and some types of flower species. Hummingbirds rely on nectar of certain flowers for nutrition. Flowers are sexual organs of plants. In a flower, the male sexual organ is the stamen. The stamen produces pollen, the plant version of sperm cells (see Figure 5.13).

To reproduce, the plant needs to get the pollen to the female sexual organ; stigma. The stigma could be part of the same plant or a different flowering tree of the same species. Like certain microbes, some plants could reproduce asexually (without a need for a mate). In such cases, the pollen just needs to travel to another part (stigma) of the same plant. This sort of thing is known as self-pollination. If the pollen travels from the stamen of one plant to the stigma of another plant, then it's called cross-pollination. Either case, the plants often need help to transport pollen. The usual delivery agents are wind, insects or birds like the hummingbird. So plants need the hummingbirds for pollen transport while hummingbirds need flowering plants to feed on their nectar. Over time evolution has caused hummingbirds to develop long slender beaks to efficiently probe the flowers for nectar. The plant on the other hand evolved specifically shaped flower petals to place pollen on certain parts of the hummingbird's body. This pollen placement technique maximizes transport efficiency increasing its chances of reproduction (Canela et al., 2003; Cotton, 1998). The co-evolution of the hummingbird and the flower is a win-win relationship. Unfortunately this is not always the case, the Red Queen has more wicked undertones.

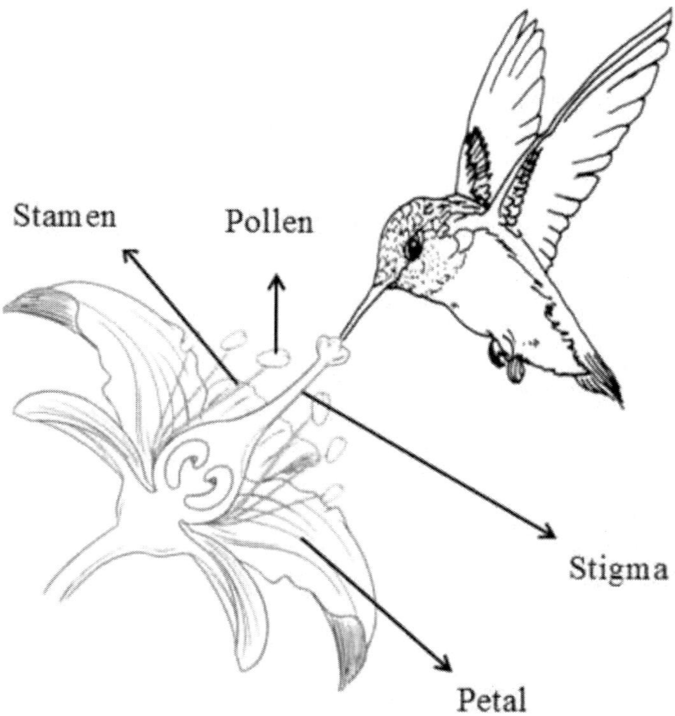

Figure 5.13. Hummingbirds and Pollination.

Some organisms tend to reap more benefits from the relationship with their partners. The Lycaenid butterfly is one such species. Lycaenid butterflies often need the help of ants for protection (Hojo et al., 2015; Pierce et al., 2002). When the butterflies are caterpillars, they secrete a sweet liquid which is consumed by the ants. So after a day of grazing, the caterpillars are escorted back to their nest safely up the trees by the ants. The ants keep other predators away in return for the sweet secretion produced by the caterpillars. Though this looks like a win-win, the caterpillars gain more from the relationship compared to the ants. The ants only get a little additional nutrition from the caterpillar. It still requires other food sources to ensure its survival. But the caterpillar gets direct protection from threats of predators thanks to the ants. Since the caterpillar's continual survival is heavily dependent on the ant but not the reverse, the relationship is a tad one-sided.

The Red Queen Effect takes an even more aggressive turn when it comes to predator-prey relationships. Take for instance the rough-skinned newt which is native to Northern America. When under threat, the newt warns by giving off an acrid smell. The skin of the newt produces a very potent neurotoxin making it lethal if consumed (even to humans). The newt's toxin is targeted towards its natural predator, the garter snake. But garter snakes evolved resistance towards the newt's toxin. Reacting to this the newt evolved by increasing the toxicity levels of its secretion. This way each organism evolves with respect to each other maintaining their predator-prey relationship. Overreaction by any one of the species may tilt the scales towards one party. And this may have grave consequences, possibly resulting in extinction of one species. Besides chemical toxins, the Red Queen Effect also boosts the evolution of senses. Previously we talked about how bats use sonar (sound waves) to echolocate. Some bats prey on the greater wax moth. By using echolocation, bats bounce of sound waves on walls of caves to track moths. Different animals have different sensitivity towards sound. Due to this difference, the frequency range of sound that could be perceived varies depending on the animal. Humans typically could hear sound waves with the frequency of 20 to 20,000 Hz. Above that frequency, we hear nothing. Bats on the other hand can detect sound waves up to a frequency of 100,000 Hz. So if a prey does not have that kind of hearing sensitivity, they wouldn't know if they are being echolocated by nearby bats. This way bats could easily sneak-up on them making them instant meals. This is when the Red Queen Effect steps in – the moths then start evolving hearing that could detect higher frequency sounds. With this, the moths could now escape the bats. Now the bats evolve back pushing up its echolocation to a frequency of 200,000 Hz. A recent research done by Hannah Moir and her team from the University of Strathclyde, Glasgow showed that some moths could detect sound waves up to a staggering 300,000 Hz (Moir et al., 2013). The interesting thing is that above 200,000 Hz, sound doesn't travel well. At such high frequencies, sound dissipates and gets heavily interfered while propagating in air. This is why bats don't echolocate at such high frequencies. The arms-race between the moths and bats have somewhat reached the limits of the physics of sound. How about that. This is the strength of the Red Queen Effect (and evolution in general). A clear example of evolution pushing the ability of an organism so far that it reaches a physical constraint. This makes one wonder about human intelligence. If via competition and survival we evolved our superior intelligence, would there be a physical limit to our intelligence (or brain structure) similar to the frequency limit faced by the bats? How much more potential is left for our intelligence to evolve?

WHEELS WITHIN WHEELS

The most riveting (and mind-boggling) thing about life is its sheer complexity. Not only whole creatures are complicated, even their individual parts are often sophisticated. Scientists still couldn't understand how some organs got to their current states – their evolutionary history. With an extreme level of intricacy, the human eye stands out among the many complex structures in nature. A human eye is composed of multiple functioning parts, all working in tandem much like the insides of a mechanical clock. First there is the retina, which refracts light from the outside and focuses it on the retina. The retina is a component at the back of the eyeball. Its sensitivity enables it to convert light into electrical signals which are then sent to the brains via the optic nerve. Then we have the cornea which is the transparent layer covering the front part of the eye. Right behind the cornea there is a ring-shaped layer called the iris. This component controls the size and thus the opening of the pupil. This is to regulate the amount of light reaching the retina. And the list of components goes on reaching about thirty known parts. Each part performing its own function keeping in sync with other parts, like clockwork. Take one component out and the whole thing collapses, resulting in blindness. If we consider evolution, then the build-up of complexity is not so surprising. First you an organism perhaps with some rudimentary abilities enough for it survive. Then as times change, it faces challenges from the environment as well as other organisms. Mutations happen in the population. Those with advantageous mutations live on while the others eventually die out. So far, so good. Then this goes on for a while until new challenges arise and the process repeats. Therefore, over very lengthy periods, the organism accumulates mutations building on past mutations – steadily evolving. This would cause it to become more complex as it gathers up these changes. Not to mention if the organism splits into a different species like our Galapagos iguanas, then this would significantly add-on to the already complicated being. So using evolution, we get a rough idea of how life gets complicated. But how about at the beginning; when things were just starting? What was the source of all this complexity?

The Cambrian Explosion caused a major diversification of species on our planet. Many organisms that flourished during that period were multicellular – organisms with multiple cells as oppose to those with single cells (such as bacteria and many other microbes). But what about before that? Our best guess is that, at that time it was the age of bacteria. A time when single-celled organisms thrived on earth. Nicole King, a biologist at University of California, Berkeley attempted to find the link between the age of bacteria and the time of the Cambrian Explosion (McGowan, 2014). In other words her research centered on the question of how single-celled organisms became multicellular creatures. Looking at fossil evidence, King realized the data was ambiguous at best since not much well-preserved fossils exist before the Cambrian explosion. So she started investigating a microscopic aquatic creature, choanoflagellates. The name choano-flagellates originates from the combination of the Greek word, *Khoanē* which means 'funnel' and the Latin word *flagellum* is used to describe the long-tailed microbes. Since choanoflagellates is a mouthful, we'll just call them *choanos*. Choanos have a funnel-like shape connecting its tail (flagellum) to the rest of its oval-shaped body. The flagellum allows it swim around in the water as well as 'pull' bacteria towards it to be eaten. Consuming mostly other bacteria and organic matter, choanos are the closest living relative of animals today. King's findings indicate that bacteria were the primary food source for pre-Cambrian organisms. They thus play a pivotal role in the development of animal life on the planet. This means that the choanos are similar to some of the Cambrian animals – its genetic material could be found in almost all animals. Unfortunately only choanos are still around since most of the Cambrian creatures have long gone extinct. The strange thing about this microbe is that it can exist as a single-celled or multicellular organism. This is what caught the attention of King.

Most organisms on the planet can only exist in one of the forms above, not both (single-celled or multi-celled). Choanos were odd this way, they seemed to be able to switch forms whenever they see fit. King observed that choanos divide their cells like most animal cells do when they grow. Except that choanos don't completely split, the new cells remain connected to the initial cell by a thin filament. Repeating the process in the lab, the choanos managed to reach about 50 cells. She noticed that the process bared close resemblance to the one occurring when animals developing from a zygote to an embryo – before reaching the fetal stage. Although she observed this in a natural setting, King could not reproduce this effect in the lab. She couldn't get them to become multicellular. Keep in mind these cells are not dividing like a reproducing bacteria. What we are discussing here are cells that are undergoing some kind of metamorphosis like a caterpillar becoming a butterfly. Only that in this case the choanos are metamorphosing from an organism with a single cell to one with multiple cells. In one experiment, King finally figured it out when she noticed how the choanos reacted to a treatment of antibiotics. The antibiotics killed off one species of bacteria allowing the other to thrive. In an environment rich with this bacteria, the choanos instantly switches state – becoming multicellular. Recall that choanos eat bacteria. It seemed that a chemical compound produced by these this species of bacteria signals the choanos to become multicellular. It's as though these bacteria indicate to the choanos that the conditions are now favorable, so go on and transform. King and her colleagues speculate that this was most likely how the first multicellular species emerged millions of years ago before the Cambrian Explosion. She explains that perhaps many of the initial microbes could switch back and forth becoming single or multi-celled. And they only become permanently multicellular when the environment becomes completely stable.

All this sounds a little similar with the quantum complementary principle; where wave-particle duality of an electron collapses becoming a particle after observation. The organism could thus be multicellular (particle) or single-celled (wave) when the environment is unstable (before observation). When the environment reaches stability (observation occurs), the organism 'collapses' becoming multicellular (particle). It seems that we could easily use the complementary principle to draw analogues to biological life – Bohr, Delbrück and Jordan were clearly unto something. King's research focused on the first ancestors of all animals trying to understand its evolution. She offers an exciting insight into the time long before the first fossilized animals emerged. Her research highlights the importance of bacteria in the evolution of life on our planet. Even now we are very much dependent on these tiny creatures for our survival. Humans are mostly made out of bacteria. Bacteria outnumber human cells by a ratio of ten to one. We have bacteria on our skin, lung, saliva, mucus and guts. Bacteria help protect our cells against foreign infections, digest our food and help us perform many other bodily functions. This is why scientists like King believe that bacteria are the bedrock of life on earth. So we come back to the question of how a complex organ like the eye can form. Following the ideas of King, first the eye components may have been single-celled microbe – most likely similar to choanos. Each group of microbe being a species by itself functions in its own way. Perhaps some of them could refract and direct light like a lens while others could absorb light like the retina. Using their respective features they trapped and ate bacteria. At some point in their evolution they must have worked together trapping bacteria prey more efficiently. Long enough they co-evolved – Red Queen Effect happens. Then they shared genetic material and became a single organism. As time passes, this keeps happening, more microbes get strung together and evolution does its work. And you get an eye in the end. Or is it the end? Not necessarily, our bodies should be still evolving. There's no reason why evolution should stop since we are still adapting and interacting with other life forms on the planet – bacteria, viruses and other animals. Not to mention plants which are our food source.

Previously we discussed how the mitochondrion acts as the cell's engine converting sugars to chemical fuel (ATP) to be burned for energy. It has been known that mitochondria in cells have their own RNA different from the cell's DNA or RNA (Martin et al., 2015). Thus there have been theories stating that the mitochondria may have been initially a separate organism. Then after some time it started a symbiotic relationship with a cell. This would be a win-win situation since the mitochondria could provide fuel to the cell while the cell could keep it alive by providing it with nutrients. And co-evolution happened and next thing you know it has become part of the cell. After all, there have been instances where symbiotic organisms have been found in the cells of microbes. One example is the single-celled microbe called *Paramecium Bursaria* (Karakashian, 1974). Residing within its cell, this microbe has another microbe, a type of green algae called *zoochlorellae*. This alga produces food via photosynthesis. It's likely that the *Paramecium* offers protection and some nutrients to the algae while the algae in return photosynthesizes food for the *Paramecium*. Perhaps in millions of years the algae would be absorbed into the cell of the *Paramecium* as has happened to the mitochondrion. It would then share genetic material and become a part of the *Paramecium*.

ALL THIS AND BRAINS TOO

Human-like apes appeared on the planet approximately 6 to 8 million years ago. Then suddenly at about 3 million years ago the brain underwent massive expansion. Remember at these timelines the Homo sapiens species (humans) have not yet come along. Thus we are talking about human-like apes (hominins) from the order of Hominidae (refer to Figure 5.10). Before the last 3 million years, the brains of hominins didn't grow much as compared to its ape cousins and past ancestors (Jabr, 2015). This sudden change in the past 3 million years ago is known as the Brain Boom. Paleontologists identified this period by analyzing, dating and comparing skulls of remains. The first Homo sapiens appeared about 200,000 year ago, just in time for them to receive a brain almost 1 kilogram heavier than their hominin ancestors. So for not very clear reasons, the Brain Boom started during the time of the hominins. They seemed to be growing the brain matter required for serious thinking. Then Homo sapiens came along and continued the boom working with already sophisticated hardware; thanks to the hominins. In the previous chapter we discussed how humans have way more neurons (86 billion) compared to creatures such as frogs, Guinea pigs and ants.

But then intelligence is not all about neuron numbers and brain size. Architecture is important too. If you have many neurons in the section of the brain that doesn't do much thinking, then intelligence wouldn't grow. The human brains could be broadly divided into two sections: the cerebellum and the cerebral cortex. The cerebellum part is used to regulate bodily functions like the circadian rhythm, blood pressure and heart rate. Complex and abstract thinking, language, self-awareness and art are done by the neurons in the cerebral cortex. An average human has 69 billion neurons in the cerebellum and 16 billion in the cerebral cortex. Elephants on the other hand have 251 billion neurons in the cerebellum and 5.6 billion in the cerebral cortex. Humans comparatively have about 3 times more neurons in the cerebral cortex. This is why although elephants have more neurons in total, the neurons can't be utilized for much thinking because they are not in the thinking section of the brain (cerebral cortex).

But this might not be a bad deal for an elephant since it needs to allocate lots of brain power for the functioning of its massive body. A closer comparison would be the brains of orangutans, gorillas and chimps. The cerebral cortices of gorillas and orangutans have 9 billion neurons while chimps have 6 billion. Bottom line is; getting the neurons to be in the right place is pretty important in the intelligence game.

Suzana Herculano-Houzel, a neuroscientist at the Federal University of Rio de Janeiro does research focusing on the evolution of primate brains. Her findings indicate that the stark difference between humans and other animals is the neural density in the cerebral cortex (Herculano-Houzel, 2009). Taking up only 2 percent of the overall body weight, the human brain consumes 20 percent of the body's energy. This means that 20 percent of all food we consume is used to just power the brain. Thus it stands as the most energy intensive organ in the human body. Interrupting this supply of energy could lead to neurological problems such as a stroke or spinal cord injury.

The way our brain evolved may be likened to the tongue evolution of the Hydromantes salamander. This salamander could launch its tongue to catch prey – similar to frogs and chameleons. Launching its tongue in less than 5 milliseconds, it beats most of the other tongue-launchers in the animal kingdom (Deban et al., 1997; Debam et al., 2007). Even chameleons take 20 milliseconds to shoot out its tongue. Interestingly, nothing else about the salamander is fast. And this evolutionary quirk appears independently in three different species of salamanders (perhaps a case of convergent evolution). Biologists realized that unlike other salamanders, Hydromantes salamanders have no lungs. To acquire a high-speed ballistic tongue, these salamanders had to trade-off their lungs. This is because to do this the tongue requires additional muscle power. This additional power is taken from the muscles usually used to drive the lungs. Being stronger and larger, this muscle springs around the bone at the back of the mouth. So when it launches its tongue, part of its bone dislodges as well moving out of its mouth. The Hydromantes salamander had to compromise its lungs for gaining additional tongue-launching power. This is sort of an evolutionary trade-off where nature allocates and efficiently prioritizes its resources. In the case of the salamander, catching prey (food) clearly takes precedence compared to breathing with lungs. Anyway lungs are redundant for salamanders, since most of them usually breathe with gills like fish. So out goes the lungs and in comes crazy quick tongue-launching ability.

How about humans? What did we lose to gain additional brain power? According to research by biologist Kasia Bozek, we lost muscle strength. Working at the Okinawa Institute for Science and Technology, Bozek examined the chemical aspect of the problem. She studied the concentrations of sugars and other chemicals crucial for muscle function. This way she linked chemical concentrations to energy requirements of muscles – which directly translates to muscle strength. Closely related species like chimps are very similar to us in this respect. Bozek then carried out an experiment consisting of human and simian test subjects. The human test subjects included professional rock-climbers and basketball players. Their simian counterparts consisted of macaques and chimps. The experimental results showed that our simian cousins were twice as strong as us. Their muscles consumed more power and similarly exerted more force compared to humans. Bozek's findings support the notion that perhaps at the expense of intelligence and brain power, unlike our simian cousins we lost much of our muscle strength.

In the same spirit, anthropologist, Leslie Aiello and biologist, Peter Wheeler came up with the 'Expensive Tissue Hypothesis' (Aiello and Wheeler, 1995; Aiello, 1997). Just as the salamander traded-off its lungs for a faster tongue, the human gut shrank to support a larger energy-intensive brain. And this shrinking is in addition to the loss in overall muscle strength. This way the energy previously used by the intestines could be directed to the brain. They built their idea by comparing data on brain and gut sizes of humans against those of primates. Richard Wrangham, a primate specialist studied the digestive system of a python (Carmody and Wrangham, 2009). He showed that the python used less energy to digest cooked food. It takes less effort to break it down compared to eating it raw. This is because heat alters and preprocesses the chemical compounds in food. This is why cooked food needs less effort to breakdown – and thus consumes less energy to digest. Seeking further evidence, Wrangham studied modern hunter-gatherer tribes around the world (Gorman, 2008). He found that none of them actively consume raw food – this means that almost all modern humans are completely adapted to eat cooked food. Wrangham's calculations show that the amount of calories that we gain from raw meat is considerably less than cooked food. So if we decide to eat what chimps eat daily, we most probably would not last long – due to the low calorific value of raw chimp food. Studies have shown that women on a raw vegetarian diet have disturbances in their menstrual cycle. This is probably caused by an energy deficit resulting from eating raw food – making it difficult for the body to run its regular cycles.

Aside from cooking, biologists have also connected dietary changes to rapid brain evolution. Curtis W. Marean, the associate director of Arizona State University's Institute of Human Origins has been involved in the excavations at Pinnacle Point, a site in South Africa. His research indicates that humans had a drastic change in diet about 160,000 years ago (Marean, 2014). Instead of foraging for plants and hunting animals, they started eating shellfish and other kinds of marine life. High in omega-3 fatty acids and other nutrients (iron, zinc, copper, iodine and selenium), marine animals could have significantly aided in brain growth of humans back then. Besides fish is a much more reliable food source. With this stability, prehistoric humans would have had more time and security to develop their cognitive abilities (e.g., culture, art and sophisticated tools). Social networks are another factor that may have stimulated brain evolution. Communal life would have made looking for resources such as food and shelter more efficient and easier. In addition, to effectively communicate, extensive language development would have been essential. And language development is the key ingredient for accelerating intelligence. Sudden changes in the prehistoric climate could have also contributed (Reed, 1997). Changes in climate would have directly affected resources needed for the survival of humans. Thus, relocation would have become inevitable. Relocation to regions with near wetlands and the ocean would have made them adopt an omega-rich diet of marine animals. If we add cooking into all this, then brain growth would naturally come along. It is highly probable that all of the factors occurred together (perhaps not in the order described above) and thus affected the brain development of prehistoric man.

Having a really smart brain though may not be enough. Dolphins and wolves are quite intelligent. Let's not forget parrots, which can imitate our use of language. But they could hardly manipulate their surroundings and have control of their resources as we do. Their intelligence is somewhat limited by their physiology.

Humans are lucky enough to have a physiology that is cohesive to the development of our intelligence. Thanks to simian origins, we have prominent thumbs which could be used for building complex tools. Besides by standing upright (bipedal), we freed our hands to do other stuff. Not to mention our vocal tract which we use to communicate. This catalyzes the development of language which speeds brain evolution.

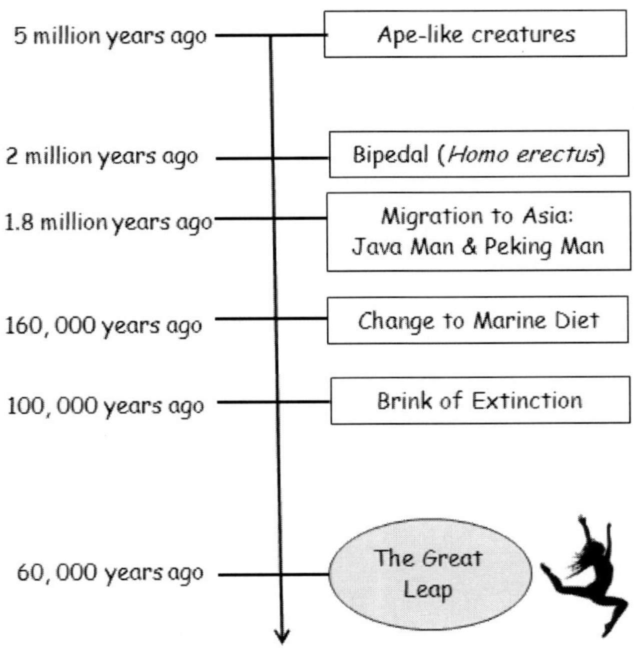

Figure 5.14. Hummingbirds and Pollination.

The interesting thing is that intelligence did not immediately come with brain development. The Brain Boom happened 3 million years ago causing the beginning of brain growth (Figure 5.14). After 1 million years, bipedal apes started appearing (ancestors to humans called *Homo erectus*). Fossils of *Homo erectus* were found in Africa. Then these apes started moving out of Africa to Asia. The Java Man and Peking Man fossils found in Asia are evidence of this migration. The migration happened about 1.8 million years ago. Homo sapiens emerged at about 200, 000 years back and went through a dietary transition (about 20,000 years after). Then the numbers dropped so low that during 100, 000 years back, the whole species was in on the verge of extinction. There's a period when fossil evidence indicate nothing happened – no difference in tools used and the cave paintings were identical to those found 200, 000 years back. But during this period brain size did increase. Then suddenly at about 60, 000 years back, sudden signs of intelligence seem to appear in humans. Excavation sites during this period seem to contain weapons, finely crafted tools and even more intricate cave paintings (art). There are even evidence of the use of body ornaments, funeral ceremonies and jewels in the remains. This sudden birth of intelligence is known as 'The Great Leap'. It was as though the brain was growing and developing the hardware. And as it kept on doing this it was reaching a new attractor (like described in Figure 4.4. in the previous chapter). Once the brain matter was ready, it shifted crossing a critical line into a new attractor – a new state of intelligence. It may seem that development of intelligence in humans is similar to the behavior of complex systems. The Great Leap signifies the transition to a higher state of intelligence. Which makes one wonder, are there anymore such transitions? Would the human mind evolve again towards a new attractor – 'leaping' to an even higher state of intelligence?

IN BRIEF

Like many current technologies, biology and medicine has gone into the quantum age. Using quantum physics, we can now unravel some of the most puzzling natural phenomena like plant photosynthesis, bird migration and genetic mutation. Fossil records at the Burgess shale show that life on the planet was suddenly diversified in an event called the 'Cambrian Explosion'. We also explore how things would have been if the conditions were different during the Cambrian Explosion – especially with respect to our own evolution. Evolution is seen to be powered by two main drivers: convergence and contingency. To survive, organisms tend to co-evolve with others giving rise to the Red Queen Effect. Microbes like the choano explain how complex life initially formed even before the Cambrian Explosion. After appearing about 200,000 years ago, Homo sapiens had brains much heavier than their hominin ancestors. To accommodate this large brain, Homo sapiens traded-off overall muscle strength and digestive capability (which resulted in a weaker gut). In addition to cooking, tools were used to compensate for these losses. Although having large brains, their intelligence remained dormant – not changing much for a few hundred thousand years. Then suddenly, about 60,000 years back, records indicate a sudden burst of intelligence among Homo sapiens; known as the Great Leap.

REFERENCES

Adler, J., (2013). How the Burgess Shale Changed Our View of Evolution: The famed fossils are a link to some of the first complex creatures on Earth, *Smithsonian Magazine*: http://www.smithsonianmag.com/history/how-the-burgess-shale-changed-our-view-of-evolution-3678444/

Aiello, L.C. and Wheeler, P., 1995. The expensive-tissue hypothesis: the brain and the digestive system in human and primate evolution. *Current anthropology,* 36(2), pp.199-221.

Aiello, L.C., 1997. Brains and guts in human evolution: the expensive tissue hypothesis. *Brazilian Journal of Genetics,* 20.

Arndt, M., Juffmann, T. and Vedral, V., 2009. Quantum physics meets biology. *HFSP journal,* 3(6), pp.386-400.

Arnold, C., (2015). Ants Swarm Like Brains Think: A neuroscientist studies ant colonies to understand feedback in the brain, *Nautilus*, no 23: http://nautil.us/issue/23/dominoes/ants-swarm-like-brains-think-rp.

Barr, S.I., 1999. Vegetarianism and menstrual cycle disturbances: is there an association?. *The American journal of clinical nutrition,* 70(3), pp.549S-554S.

Bennett, K., (2010). The chaos theory of evolution, *New Scientist*: https://www.newscientist.com/article/mg20827821-000-the-chaos-theory-of-evolution/

Bohr, Niels Henrik David, Complete Dictionary of Scientific Biography. *Encyclopedia.com.* (February 17, 2017). http://www.encyclopedia.com/science/dictionaries-thesauruses-pictures-and-press-releases/bohr-niels-henrik-david.

Born, M.; Heisenberg, W.; Jordan, P. (1926). "Zur Quantenmechanik. II." . [To quantum mechanics. II.] *Zeitschrift für Physik* 35 (8–9): 557. Bibcode:1926 ZPhy...35..557B. doi:10.1007/BF01379806.

Brooks, M., (2011). Quantum life: The weirdness inside us: Ideas from the stranger side of physics could explain some long-standing mysteries of biology, *New Scientist*: https://www.newscientist.com/article/mg21128321-500-quantum-life-the-weirdness-inside-us/

Canela, M.B.F. and Sazima, M., 2003. Aechmea pectinata: a Hummingbird-dependent Bromeliad with Inconspicuous Flowers from the Rainforest in South-eastern Brazil. *Annals of Botany,* 92(5), pp.731-737.

Carmody, R.N. and Wrangham, R.W., 2009. The energetic significance of cooking. *Journal of Human Evolution,* 57(4), pp.379-391.

Cartlidge, E., 2010. Quantum mechanics boosts photosynthesis, IOP *Physics World*: http://physicsworld.com/cws/article/news/2010/feb/04/quantum-mechanics-boosts-photosynthesis.

Collini, E., Wong, C.Y., Wilk, K.E., Curmi, P.M., Brumer, P. and Scholes, G.D., 2010. Coherently wired light-harvesting in photosynthetic marine algae at ambient temperature. *Nature,* 463(7281), pp.644-647.

Cotton, P.A., 1998. Coevolution in an Amazonian hummingbird-plant community. *Ibis*, 140(4), pp.639-646.

Courtland, R., (2011). Fly sniffs molecule's quantum vibration, Daily News, *New Scientist*: https://www.newscientist.com/article/dn20130-fly-sniffs-molecules-quantum-vibrations/

Dawkins, R. and McKean, D., 2012. *The Illustrated Magic of Reality: How We Know What's Really True.* Simon and Schuster.

Deban, S.M., Wake, D.B. and Roth, G., 1997. Salamander with a ballistic tongue. *Nature,* 389(6646), pp.27-28.

Deban, S.M., O'Reilly, J.C., Dicke, U. and Van Leeuwen, J.L., 2007. Extremely high-power tongue projection in plethodontid salamanders. *Journal of Experimental Biology,* 210(4), pp.655-667.

Dell'Amore, C., (2013). Giant Platypus Found, Shakes Up Evolutionary Tree: Three-foot mammal lived about 5 to 15 million years ago, *National Geographic*: http://news.nationalgeographic.com/news/2013/11/131104-giant-platypus-evolution-science-animals-paleontology/

Diamond, J., (1992). *The Rise and Fall of the Third Chimpanzee,* Vintage. ISBN 0099913801.

Engel, G.S., Calhoun, T.R., Read, E.L., Ahn, T.K., Mančal, T., Cheng, Y.C., Blankenship, R.E. and Fleming, G.R., 2007. Evidence for wavelike energy transfer through quantum coherence in photosynthetic systems. *Nature,* 446(7137), pp.782-786.

Fallon, John F.; Simandl, B. Kay (1978). "Evidence of a role for cell death in the disappearance of the embryonic human tail." *American Journal of Anatomy*. 152 (1): 111–29.

Fischer, E. P, (2007), Max Delbrück, *Genetics* October 1, 2007 vol. 177 no. 2, 673-676.

Foster, P.L., 2004. Rebuttal: adaptive point mutation (Rosenberg and Hastings). *Journal of bacteriology, 186*(15), pp.4845-4845.

Foley, L.E., Gegear, R.J. and Reppert, S.M., 2011. Human cryptochrome exhibits light-dependent magnetosensitivity. *Nature communications,* 2, p.356.

Franco, M.I., Turin, L., Mershin, A. and Skoulakis, E.M., 2011. Molecular vibration-sensing component in Drosophila melanogaster olfaction.

Proceedings of the National Academy of Sciences, 108(9), pp.3797-3802.

Gorman, R.M., 2008. *Cooking Up Bigger Brains,* Evolution, *Scientific American*: https://www.scientificamerican.com/article/cooking-up-bigger-brains/

Gould, S.J., 1990. *Wonderful Life: The Burgess Shale and the Nature of History,* Vintage Books.

Grossman, L., (2011). In the Blink of Bird's Eye, a Model for Quantum Navigation, *Science,* Books.

Herculano-Houzel, S., 2009. The human brain in numbers: a linearly scaled-up primate brain. *Frontiers in human neuroscience,* 3, p.31.

Hoffmann, F.G., Opazo, J.C. and Storz, J.F., 2010. Gene cooption and convergent evolution of oxygen transport hemoglobins in jawed and jawless vertebrates. *Proceedings of the National Academy of Sciences,* 107(32), pp.14274-14279.

Horovitz, I., Martin, T., Bloch, J., Ladevèze, S., Kurz, C. and Sánchez-Villagra, M.R., 2009. Cranial anatomy of the earliest marsupials and the origin of opossums. *PLoS One,* 4(12), p.e8278.

Hunter, R.J. (1989). *Foundations of Colloid Science.* Oxford University Press.

Jabr, F., (2015). How Humans Evolved Supersize Brains: Scientists have begun to identify the symphony of biological triggers that powered the extraordinary expansion of the human brain. *Quanta Magazine*: https://www.quantamagazine. org/20151110-evolution-of-big-brains/

Joaquim, L., Freire, O. and El-Hani, C.N., 2015. Quantum Explorers: Bohr, Jordan, and Delbrück Venturing into Biology. *Physics in Perspective,* 17(3), pp.236-250.

Karakashian, M.W., 1974, December. Symbiosis in Paramecium Bursaria. In *Symposia of the Society for Experimental Biolog*y (No. 29, pp. 145-173).

Klein, R. and Edgar, B., (2002). *The Dawn of Human Culture,* John Wiley. ISBN 0471252522.

Kryazhimskiy, S., Rice, D.P., Jerison, E.R. and Desai, M.M., 2014. Global epistasis makes adaptation predictable despite sequence-level stochasticity. *Science,* 344(6191), pp.1519-1522.

Langley, L., (2015). Which Animals Have Barely Evolved?: Some species have managed to maintain their original looks—even over millions of years, *National Geographic*: http://news.nationalgeographic.com/2015/11/151114-animals-mammals-evolution-platypus-opossum/

Langley, L., (2016). How the Venomous, Egg-Laying Platypus Evolved: The odd Australian mammal has an intriguing family tree, *National Geographic*: http://news.nationalgeographic.com/2016/07/animals-platypus-evolution-science/

Lister, A.M., 2013. The role of behaviour in adaptive morphological evolution of African proboscideans. *Nature,* 500(7462), pp.331-334.

Löwdin, P.O., 1963. Proton tunneling in DNA and its biological implications. *Reviews of Modern Physics,* 35(3), p.724.

Lyklema, J. (1995). *Fundamentals of Interface and Colloid Science.* vol. 2. p. 3.208.

Marean, C.W., 2014. The origins and significance of coastal resource use in Africa and Western Eurasia. *Journal of human evolution,* 77, pp.17-40.

Martin, W.F., Garg, S. and Zimorski, V., 2015. Endosymbiotic theories for eukaryote origin. *Phil. Trans. R. Soc. B,* 370(1678), p.20140330.

Matthews, S.E., (2013). 5 Animals With An Extraordinary Sense Of Smell Moths, sharks, fish, and more, *Popular Science*: http://www.popsci.com/science/article/2013-05/five-feats-smell.

McFadden, J., 2017. http://www.johnjoemcfadden.com/popular-science/quantum-biology/coherence-in-photosynthesis/

McGowan, K., (2014). How Life Made the Leap From Single Cells to Multicellular Animals, Wired (originally appeared in *Science, Quanta Magazine*): https://www.wired.com/2014/08/where-animals-come-from/

Merali, Z., (2014). Solving Biology's Mysteries Using Quantum Mechanics, *Discover*, http://discovermagazine.com/2014/dec/17-this-quantum-life.

Meredith, P., (2014). *How birds use Earth's magnetic field to navigate, Australian Geographic*: http://www.australiangeographic.com.au/topics/science-environment/2014/05/animal-magnetism-how-animals-use-earths-magnetic-field.

Moir H.M., Jackson J.C. & Windmill J.F.C. (2013). Extremely high frequency sensitivity in a 'simple' ear, *Biology Letters,* 9 (4) 20130241-20130241. DOI: 10.1098/rsbl.2013.0241.

Ofria, C. and Wilke, C.O., 2005. Avida: Evolution experiments with self-replicating computer programs. In *Artificial Life Models in Software* (pp. 3-35). Springer London.

Packer, M.S. and Liu, D.R., 2015. Methods for the directed evolution of proteins. *Nature Reviews Genetics,* 16(7), pp.379-394.

Parker, J., Tsagkogeorga, G., Cotton, J.A., Liu, Y., Provero, P., Stupka, E. and Rossiter, S.J., 2013. Genome-wide signatures of convergent evolution in echolocating mammals. *Nature,* 502(7470), pp.228-231.

Pian, R., Archer, M. and Hand, S.J., 2013. A new, giant platypus, Obdurodon tharalkooschild, sp. nov. (Monotremata, Ornithorhynchidae), from the Riversleigh World Heritage Area, Australia. *Journal of Vertebrate Paleontology,* 33(6), pp.1255-1259.

Reed, K.E., 1997. Early hominid evolution and ecological change through the African Plio-Pleistocene. *Journal of human evolution,* 32(2-3), pp.289-322.

Ritz, T., Adem, S. and Schulten, K., 2000. A model for photoreceptor-based magnetoreception in birds. *Biophysical journal,* 78(2), pp.707-718.

Roberts, A., (2016). What an elephant's tooth teaches us about evolution, *The Guardian*: https://www.theguardian.com/science/2016/jan/31/elephant-teeth-teach-about-evolution

Rosenberg, S.M., Shee, C., Frisch, R.L. and Hastings, P.J., 2012. Stress-induced mutation via DNA breaks in Escherichia coli: A molecular mechanism with implications for evolution and medicine. *Bioessays,* 34(10), pp.885-892.

Saarinen, J., Karme, A., Cerling, T., Uno, K., Säilä, L., Kasiki, S., Ngene, S., Obari, T., Mbua, E., Manthi, F.K. and Fortelius, M., 2015. A New

Tooth Wear–Based Dietary Analysis Method for Proboscidea (Mammalia). *Journal of Vertebrate Paleontology*, 35(3), p.e918546.

Sautrey, G., Duval, R.E., Chevalley, A., Fontanay, S. and Clarot, I., 2015. Capillary electrophoresis for fast detection of heterogeneous population in colistin-resistant Gram-negative bacteria. *Electrophoresis,* 36(20), pp.2630-2633.

Schlau-Cohen, G.S., Ishizaki, A., Calhoun, T.R., Ginsberg, N.S., Ballottari, M., Bassi, R. and Fleming, G.R., 2012. Elucidation of the timescales and origins of quantum electronic coherence in LHCII. *Nature chemistry,* 4(5), pp.389-395.

Schrödinger, E., 1944 *What is Life*? Cambridge University Press, Cambridge, UK.

Singer, E., (2014). Under Pressure, Does Evolution Evolve?, Evolution, *Quanta Magazine*: https://www.quantamagazine.org/20140115-under-pressure-does-evolution- evolve/

Sniegowski, P.D., Gerrish, P.J. and Lenski, R.E., 1997. Evolution of high mutation rates in experimental populations of *E. coli. Nature,* 387(6634), p.703.

Well, S., (2003). *The Journey of Man: A Genetic Odyssey. Penguin.* ISBN 0141008326.

Whittington, C.M., Papenfuss, A.T., Locke, D.P., Mardis, E.R., Wilson, R.K., Abubucker, S., Mitreva, M., Wong, E.S., Hsu, A.L., Kuchel, P.W. and Belov, K., 2010. Novel venom gene discovery in the platypus. *Genome biology,* 11(9), p.R95.

Wiltschko, W. and Wiltschko, R., 2005. Magnetic orientation and magnetoreception in birds and other animals. *Journal of Comparative Physiology A,* 191(8), pp.675-693.

Zimmer, C., (2010). 7,000 Miles Nonstop, and No Pretzels, Science, *The New York Times*: http://www.nytimes.com/2010/05/25/science/ 25migrate.html.

Zorich, Z., (2016). 'If the World Began Again, Would Life as We Know It Exist?: Experiments in evolution are exploring what would happen if we rewound the tape of life, *Nautilus*: http://nautil.us/issue/

34/adaptation/if-the-world-began-again-would-life-as-we-know-it-exist-rp.

Websites

Palaeontology[online]: Palaeontology on the Web (2017), http://www.Palaeontologyonline.com/articles/2013/patterns-in-palaeontology-old-shapes-new-tricks-the-study-of-fossil-morphology/verity_shape_fig_1/?doing_wp_cron=1489469815.5307118892669677734375.

Final Remarks

"Who looks outside, dreams. Who looks inside, awakens."
– Carl Gustav Jung

Perhaps the most monumental achievement in biology and medicine is the discovery of the DNA by Watson and Crick. It solidified the ideas presented by Schrodinger and others after him; that life is clearly driven by code, similar to the workings of computers. But life is far more complex than mere computers. The amount of processing done by a simple organism like an insect beats some of the most high-performance computers we have today. One reason life is so robust and efficient is because it does not just operate using conventional (or classical) physics – it exploits quantum physics. Quantum physics allows life to pack a lot of complexity in a small and compact form. This is why even simple microbes like choanos and *E.coli* are extremely adaptive to their environments. And they can adapt very rapidly by mutating their genes; changing their nature in a blink of an eye. Using quantum physics, the mutations are catered exactly to the environment such that when stressed, they can immediately change their diets to fit almost any available food source. Quantum physics works hand in hand with gene codes to sustain life. Mutation also evolves organisms as a reaction to the evolution of others – Red Queen Effect.

Even the way parasites evolve to mimic and cloak their sinister plots is simply astonishing. The Red Queen Effect is the foundation of co-evolution which allows them to evolve resistance to increase their chances of survival. Evolution works broadly, transforming entire ecosystems influencing chains of organisms at one go. This is why the sheer complexity of parasites can be mind-boggling. Their true function in nature continues to elude even the best scientific minds. Some parasites like the Bd fungus are so damaging they could wipe out entire species'. Nevertheless understanding these parasites and their roles in nature is crucial for medicine. We have indeed come a long way from the time of Jenner and Pasteur, in batting diseases. However, as disease-causing agents keep evolving, we find ourselves always in constant battle with them. Knowing the origins of disease would tell us how to deal with them giving us a better fighting chance against epidemics and worst still, pandemics like Zika. And the only way to do this is through science. This is why having an effective scientific system is essential. Things like publication bias and other issues that hamper the progress of science need to be ironed out. The economic system should be adjusted such that we can mobilize resources swiftly to curb epidemics before it becomes irreversible. By failing to channel our efforts this way, our species will suffer serious losses the next time we face an aggressive epidemic. We have to keep in mind that as science progresses, the way we do science has to change along with it. We need to keep tweaking it and maintaining its framework so that it doesn't deteriorate with time. This is so that we can keep making discoveries and effectively apply science whenever needed. In other words, we have to keep the scientific enterprise in good condition – and this takes work.

On a brighter note, our capability in gene editing has drastically improved with the advent of the CRISPR technique. The potential for this technique cannot be understated. Although still at its infancies, gene editing would in my opinion eventually revolutionize medicine in addition to accelerating biological research. Albeit its triumphs, we would still have to jump many ethical hurdles before we can put it to good use. Like other technologies in the past, these issues would gradually resolve as we encounter them, one at a time – assuming we keep doing good science. With the advent of Artificial Intelligence, tools used to explore biological phenomena have greatly advanced. Words like 'neural networks' and 'machine learning' are now frequently used by biologists and physicians during scientific discourse. Being a recent field like gene editing, artificial intelligence is just beginning. Even then, its potential is seen to be far-reaching.

Besides, science has matured to such an extent that we are now getting closer to understanding the origins of life on earth. And by doing this, we also understand the origins of ourselves – the beginning of our own species, Homo sapiens. Looking back at how it all began gives us an idea of how we evolved to our current selves. We now understand that although our brain size took millions of years to grow, our intelligence evolved abruptly – the Great Leap. Let's hope that is not the last time we 'leap'. Perhaps science could advance us as a species and take us far enough for us to 'leap' again.

AUTHOR CONTACT INFORMATION

Timothy Ganesan, PhD
Current: Research Engineer at TNB Research, Malaysia
Previous: University of Adelaide, Australia &
Universiti Teknologi Petronas, Malaysia
R33, Jalan Umbun, Taman Setapak, 53000, Kuala Lumpur, Malaysia
Telephone: +6001117693991
tim.ganesan@gmail.com

INDEX

C

D

E

F

G

H

I

J

K

L

M

W

X

Y

Z